3rd International Conference on Green Energy and Environment Engineering (CGEEE 2020)

Held online due to COVID-19

IOP Conference Series: Earth and Environmental Science
Volume 633

Jeju Island, South Korea
22 – 24 July 2020

ISBN: 978-1-7138-2946-1
ISSN: 1755-1307

Printed from e-media with permission by:

Curran Associates, Inc.
57 Morehouse Lane
Red Hook, NY 12571

CURRAN ASSOCIATES INC. proceedings .com

Some format issues inherent in the e-media version may also appear in this print version.

Institute of Physics
Dirac House, Temple Back
Bristol BS1 6BE UK

Phone: 44 1 17 929 7481
Fax: 44 1 17 920 0979

techtracking@iop.org

Additional copies of this publication are available from:

Curran Associates, Inc.
57 Morehouse Lane
Red Hook, NY 12571 USA
Phone: 845-758-0400
Fax: 845-758-2633
Email: curran@proceedings.com
Web: www.proceedings.com

TABLE OF CONTENTS

CLEAN ENERGY TECHNOLOGY AND POWER GENERATION

POWER ENGINEERING AND ENERGY CONSERVATION TECHNOLOGY

RESOURCE MONITORING AND ENVIRONMENTAL ANALYSIS

IOP Conf. Series: Earth and Environmental Science **633** (2021) 011001 doi:10.1088/1755-1315/633/1/011001

PREFACE

CGEEE 2020 was organized by Hong Kong Chemical, Biological & Environmental Engineering Society (HKCBEES) and Environment and Agriculture Society (EAS). Due to the spread of covid-19 in the world, the epidemic situation in various countries is complicated, and the entry-exit management of the local government is strict. Most authors chose the online presentation. Even if the conference was postponed, there were still many unstable factors affecting the conference. Considering the safety of the authors and facilitating more authors to participate in the meeting to achieve better communication effect, conference committee finally decided to hold the 2020 3rd International Conference on Green Energy and Environment Engineering (CGEEE 2020) as virtual conference during July 22-24, 2020, which was a great success.

CGEEE 2020 is one of the leading international conferences for presenting novel and fundamental advances in the fields of green energy and environment engineering. It also serves to foster communication among researchers and practitioners working in a wide variety of scientific areas with a common interest in improving green energy and environment engineering related techniques. Prof. Jae K. Park from University of Wisconsin-Madison, USA has been invited to do the welcome address and also given his keynote speech "Distortion of Green Energy and Environmental Issues by Deliberate Abuse of Facts"; Prof. Jin keun Kim from Jeju National University, Jeju, Korea has been invited to do the opening remarks and also given his keynote speech "Smart Water Pipeline Network Management in Korea"; Prof. Yu Hong from Beijing Forestry University, China has shared her keynote speech "The Use of Microalgae for Coupling Wastewater Treatment with High-valued Products Production"; Prof. Marco Casini from SAPIENZA University of Rome, Italy has given his keynote speech "Robotics and Automated Systems in Construction: Challenges and Opportunities for the Building Industry"; Prof. Miwako Hosoda from Seisa University, Tokyo, Japan also presented her keynote speech "Condition of a Sustainable Coexistence between People and Nature".

The meeting was held online very smoothly with the tool "ZOOM", many researchers, engineers, academicians as well as industrial professionals from all over the world have presented their research results and development activities. Each author had 12 minutes for presentation and 3 minutes for question and answer. During the question time, the author can use the "Raise Hand" function or "chat" function to ask questions. There were six sessions for presentations: Wastewater Treatment, Environmental Management and Construction Science, Renewable Energy and Power Engineering, Environmental Chemistry and Groundwater Engineering, Climatology and Pollution Ecology and Agricultural Science. It is a golden opportunity for the students,

IOP Conf. Series: Earth and Environmental Science **633** (2021) 011001 doi:10.1088/1755-1315/633/1/011001

researchers and engineers to interact with the experts and specialists to get their advice or consultation on technical matters, sales and marketing strategies.

This conference proceeding presents a selection from papers submitted to the conference from universities, research institutes and industries. All of the papers were subjected to peer-review by conference committee members and international reviewers. The papers selected depended on their quality and their relevancy to the conference. The volume tends to present to the readers the advances in green energy and environment engineering and various related areas, such as energy transformation from renewable energy system (RES) to grid, artificial intelligence and machine learning studies for RESs and applications, control techniques for RESs, control technologies, grid interactive systems used in hybrid RESs, performance analysis of RESs, environmental dynamics, hydrology, meteorology, physical oceanography etc...

We would like to thank all the authors who have contributed to this volume and also to the organizing committees, reviewers, speakers, chairpersons, sponsors and all the conference participants for their support to CGEEE2020.

Prof. Jae K. Park
University of Wisconsin-Madison, USA
September 2, 2020

IOP Conf. Series: Earth and Environmental Science **633** (2021) 011002 doi:10.1088/1755-1315/633/1/011002

Conference Committee

Conference Chair

Prof. Jae K. Park, University of Wisconsin-Madison, USA

Prof. Marco Casini, Sapienza university of Rome, Italy

Program Chairs

Prof. Wallace C.H. Choy, The University of Hong Kong, Hong Kong

Prof. Jan Baeyens, Beijing University of Chemical Technology, China

Technical Committee

Prof. Marco Casini, Sapienza university of Rome, Italy

Prof. Liangnian He, Nankai University, China

Prof. Yunfa Chen, Chinese Academy of Sciences, China

Prof. dr hab. inż. Andrzej Kapłon, Kielce University of Technology, Poland

Prof. Suyin Gan, The University of Nottingham Malaysia Campus, Malaysia

Prof. Zhe Chen, Aalborg Universitet, Denmark

Prof. Young Nam Chun, Chosun University, Korea

Prof. Sunq Won Kim, Korea National University of Transportation, Korea

Prof. Michel De Paepe, Ghent University, Belgium

Prof. Keh-Chin Chang, National Cheng Kung University, Taiwan

Prof. Vinesh Thiruchelvam, Asia Pacific University of Technology and Innovation (APU), Malaysia

Prof. Zong-Xiao Yang, Henan University of Science and Technology, China

Prof. Winn-Jung Huang, Hungkuang University,Taiwan

Prof. Huang-Mu Lo,Chaoyang University of Technology, Taiwan

Prof. Jih-Hsing ChangChaoyang University of Technology, Taiwan

Prof. Ahmed A. Hossam-Eldin, Alexandria University, Egypt

Prof. Khalil El-Hami, University of Sultan Moulay Slimane, Khouribga, Morocco

Assoc. Prof. Svetla Draganova Tzvetkova, University of National and World Economy, Bulgaria

Assoc. Prof. Svenja Hanson, The University of Nottingham Malaysia Campus, Malaysia

Assoc. Prof. Nguyen Manh Khai, VNU University of Science, Vietnam

Dr. Karamanis Dimitrios, University of Patras, Greece

Dr. Almantas Pivrikas, Murdoch University, Australia

Dr. Yarusova Sofia Borisovna, Far-Eastern Branch of Russian Academy of Sciences, Russian

Dr. Ruamporn Nikhom, Thaksin University, Thailand

Dr. Ahmad Zamani Ab Halim, Universiti Malaysia Pahang, Malaysia

Dr. Raouf Ahmed Mohamed Hassan, Aswan University, Egypt

Dr. Jayant Jain, Indian Institute of Technology Delhi, India

IOP Conf. Series: Earth and Environmental Science **633** (2021) 011003 doi:10.1088/1755-1315/633/1/011003

Peer review declaration

All papers published in this volume of IOP Conference Series: Earth and Environmental Science have been peer reviewed through processes administered by the Editors. Reviews were conducted by expert referees to the professional and scientific standards expected of a proceedings journal published by IOP Publishing.

- **Type of peer review:**
  Double-blind
- **Conference submission management system:**
  Iconf submission system
- **Number of submissions received:**
  29 papers
- **Number of submissions sent for review:**
  29 papers
- **Number of submissions accepted:**
  22 papers
- **Acceptance Rate (Number of Submissions Accepted / Number of Submissions Received X 100):**
  75.8%
- **Average number of reviews per paper:**
  3 papers
- **Total number of reviewers involved:**
  18 reviewers
- **Any additional info on review process:**

1. **Preliminary Review**

   When we received authors' paper(abstract and full paper), we submitted the paper to auditing department for checking, the auditing department reviewed the content, theme, format and grammers.

2. **The full paper passed the first review would be reviewed again by conference technical committees from the following aspects: Originality, Innovation, Technical Soundness, Applicability, Clarity of presentation and Relevance.**
3. **After paper passed the first review and the second review, we sent the notification.**
4. **Paper can only be arranged for publication if it was modified according to comments and template.**

Contact person for queries:

Prof. Jae K. Park
University of Wisconsin-Madison, USA
kpark@wisc.edu

Glucose-oxygen biofuel cell based on laccase cathode and gold-modified carbon black anode: experimental research and mathematical modelling

I Arkadeva[1], V Vasilenko[1], V Bogdanovskaya[2], G Sudarev[2], E Koltsova[1]

1 Department of informational computing technologies, D. Mendeleev University of Chemical Technology of Russia, (MUCTR), Miusskaya sq., 9, 125047, Moscow, Russia
2 Laboratory of Electrocatalysis, A.N. Frumkin Institute of Physical Chemistry and Electrochemistry Russian Academy of Sciences, (IPCE RAS), Leninsky prospect, 31, 199071, Moscow, Russia
E-mail: kolts@muctr.ru

Abstract. This research studies processes in membraneless mediatorless biofuel cell (BFC), including the processes of oxygen reduction at the cathode with a laccase-based catalyst and glucose oxidation at gold-modified carbon black (20Au/XC-72R) as an anode catalyst. The influence of the glucose quantity in the nutrient solution on the electrochemical activity of BFC was investigated. At pH=4.7 the highest power density of 2.6 μW/cm^2 was obtained for a solution containing 0.5 M of glucose. It was found that the concentration of glucose in the range of 0.2-0.5 M does not affect the electrochemical activity of immobilized laccase. The mathematical model of the studied BFC was based on the continuous stirred-tank reactor model, the electrochemical processes occurring on the electrodes were taken into account. The BFC's output characteristics were calculated, profiles of the concentration of participants of electrode reactions on current density, the discharge curves, profiles of the power density dependence on current density at glucose concentration in the range of 0.1-1.5 M were obtained. The electrodes' reactions rates dependences on the current density of BFC confirm that the cathodic reaction is limiting. Optimization of the BFC operation modes showed that the optimal glucose content at the nutrition solution is 0.546 M.

1. Introduction

One of the promising directions of modern alternative energy is biofuel cells (BFC) - devices for converting the chemical energy of fuel into electrical energy using catalysts of biological nature (enzymes, microbes, etc.). BFC are alternative devices to traditional fuel cells (FC) with catalysts that contain precious metals due to some advantages: they use renewable catalysts and operate in mild conditions – the temperature range of 20-40°C and close to neutral pH, while low-temperature FCs with platinum-based catalysts function in the pH range ~1-3, at a temperature of 60-80°C and they require a proton-conducting membrane. In order to provide the manufacturing of scalable devices, it is important to develop the design of BFC without a membrane, where the participants and products of electrode reactions do not significantly affect each other's characteristics.
The interest in creating glucose-oxygen FC is mainly due to the special properties of glucose, which include: high energy intensity, low cost of production, environmental cleanliness, ease of storage, and transportation. Various approaches are used for glucose electrooxidation in BFC. First works on glucose oxidation were performed under model conditions and allowed to select the most effective

metal catalyst - a gold-modified carbon black that provides effective glucose oxidation under conditions that are favourable for the functioning of the laccase-based cathode, described in **[1]**. The use of an enzyme – glucose oxidase for glucose oxidation in BFC (or in a biosensor for glucose determination) is based on an enzymatic reaction (glucose oxidation) in the volume of the solution and oxidation on the electrode of the product of the enzymatic reaction (usually a mediator). Attempts to use enzymes: glucose oxidase or glucose dehydrogenase in the adsorbed on the electrode state for direct electrooxidation of glucose have not been successful so far.

In this paper, experimental and theoretical studies of physical and chemical processes and phenomena occurring in a membraneless mediatorless glucose-oxygen BFC with catalysts based on laccase immobilized on the surface of CNT and gold-modified carbon black were carried out. The main accent of the work was mathematical modeling of the essential functioning processes of the BFC.

2. Materials and methods

2.1. Experimental study

In this study laccase Trametes versicolor isolated by the method [2] was used. The concentration of the enzyme was 7.7 mg / ml with an activity of $0.225 \cdot 10^{-3}$ mmol ABTS$\cdot$min$^{-1}\cdot$mg^{-1}(ABTS - 2,2'-Azino-bis (3-ethylbenzothiazoline-6) sulfonic acid). Beta-D-glucose was used in studies to determine the electrochemical activity of an electrode with a catalyst based on gold-modified carbon black. Measurements were carried out in a phosphate-acetate (pH = 4~5) buffer solution prepared from $Na_2HPO_4 \cdot 2H_2O$, chemically pure acetic acid, and deionized water. The pH of the solutions was determined on the pH meter using the ECOTEST 120 ionometer from Econix. XC-72R carbon black modified with 20% gold (20Au/XC-72R), manufactured according to the method [3], was used as an anode catalyst.

The method of cyclic voltammetry (CVA) was used for registration of polarization curves at the IPC-Pro 3A potentiostat produced by the A.N. Frumkin Institute of Physical Chemistry and Electrochemistry Russian Academy of Sciences. All electrode potential measurements are made relative to a saturated silver-chloride electrode (Ag/AgCl).

The development of the design of the laboratory layout of the glucose-oxygen BFC was conducted according to the following principles:

1) Can't use expensive ion exchange membranes of the Nafion type, which acidity (pH ~ 0) does not correspond to the pH of biological systems (4.5-8);

2) Membraneless separation of the electrodes by an electrolyte chamber allows free circulation of the electrolyte and maximizes the efficiency of the electrodes active layers;

3) To increase the efficiency of oxygen supply into the reaction zone and minimize the penetration of gas bubbles into the electrolyte chamber, oxygen was supplied through the cathode chamber located on the backside of the cathode.

The scheme design of the developed BFC layout is shown in figure 1.

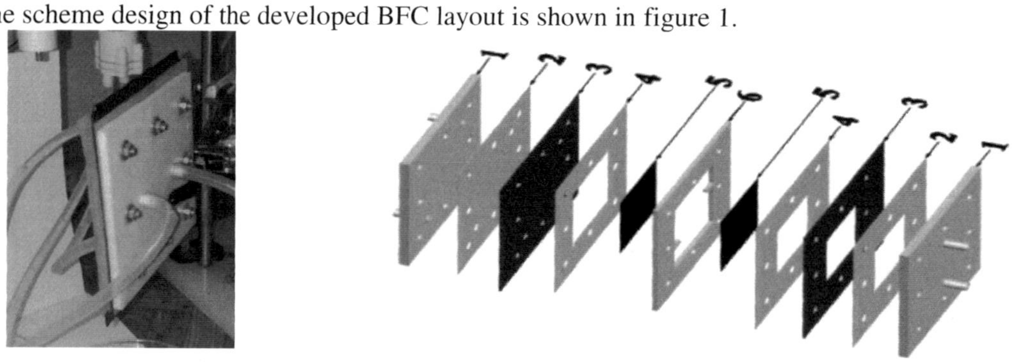

a) b)

Figure 1. a) Laboratory layout of the BFC; b) The scheme design of the BFC layout: 1 - tie plates; 2 - Teflon gaskets; 3 - current-removing plates (CRP); 4 -silicone gaskets; 5 - electrodes; 6 -electrolyte chamber.

3-D models of the tie plates and the electrolyte chamber were designed in the computer-aided design system SolidWorks CAD Premium Research and printed on a 3D Picaso Designer printer by layer-by-layer deposition of acrylonitrile butadiene styrene. The electrolyte chamber with a volume of ~ 7 cm^3 had special protrusions to prevent deformities and contact of the electrodes during cell assembly.
After preparing the cell for operation and establishing the open-circuit voltage (U_{oc}) (for ~1 h), the anode potential was measured using a silver-chloride reference electrode placed in a container at the outlet of the anode space of the cell. BFC tests were performed by measuring cyclic volt-ampere curves at a speed of 0.3 $mV \cdot s^{-1}$ from U_{oc} to U=0, as well as measuring partial curves to separate the contributions of cathode and anode processes to the total BFC voltage. To obtain partial curves, galvanostatic volt-ampere curves were measured with a holding of each point for 1 min, and then the set voltage was registered. The cathode potential for each current value was calculated by the difference of the cell voltage and the anode potential.
The electrodes for testing the BFC were formed by spraying suspensions of the catalyst and the Nafion ionomer (ionomer/carbon material ratio = 0.6) on the 39BC GDL. The loading of 20Au/XC-72R was 0.4 $mgAu \cdot cm^{-2}$. The loading of CNT was 0.4 mgC cm^{-2}. Laccase adsorbed at the electrode from its solution in the electrolyte.
The size of the electrodes was 5*5 cm^2 with an active surface area of 4*4 cm^2. The fields with a width of 5 mm at each electrode and the use of sealing gaskets provided rigidly fix the electrodes on opposite sides of the electrolyte chamber and minimize leakage of the electrolyte.

2.2. Mathematical modeling

Mathematical modeling of processes occurring in BFC is based on an approach similar to [4]. It is based on a continuous stirred-tank reactor model taking into account electrochemical phenomena. To simplify the modeling in this work, we did not consider the processes occurring in the electrode's active layers. Such modeling was considered in our previous works [5-6].
The mathematical model was developed by integrating bioelectrochemical kinetics and mass and charge balances inside the BFC. Various parameters that significantly affect the BFC's performance are studied at the simulation. Voltage, power density, and fuel concentration dependence are studied using dynamic modeling. The specified method is relatively simple and convenient to implement, which can contribute to the expansion of its practical application in the design and operation of BFCs
In accordance with the mechanism of glucose (G) oxidation on the gold catalyst described in [7], it is assumed that the anodic reaction proceeds according to the scheme (1) with a production of gluconic acid (GA), protons, and electrons.
Since the reaction rate is managed by the cell electrochemical potential Butler-Volmer expression (2) was used to describe the electrochemical reaction.

$$G + H_2O \overset{Au}{\rightarrow} GA + 2H^+ + 2\bar{e} \quad (1)$$

$$r_a = k_a \exp\left(\frac{\alpha F n_a}{RT}\eta_a\right) C_G \quad (2)$$

where C_G is the concentration of glucose in the volume of the electrolyte, $mol \cdot m^{-3}$; n_a – the number of electrons which take part in the electrode reaction; η_a is anode local surface overpotential, the difference between the electrode (anode) and electrolyte potentials, V; k_a is the rate constant of anode reaction, $m \cdot s^{-1}$; α - the charge transfer coefficient of the anodic reaction; F - the Faraday constant, $Coulombs \cdot mol^{-1}$; R is the gas constant, $J \cdot mol^{-1} \cdot K^{-1}$; T - the cell operating temperature, K. The concentration of water is assumed to be constant (an excess component in the liquid mixture).
The reduction of dissolved oxygen on the cathode is suggested to be the following (3). The rate of the cathodic oxygen reduction reaction can be described by the Monod-type equation (4).

$$L + O_2 + 4\bar{e} + 4H^+ \overset{k_c}{\rightarrow} L + 2H_2O \quad (3)$$

$$r_c = k_c \exp\left[\frac{(1-\beta)F n_c}{RT}\eta_c\right] \frac{C_{O_2}}{K_{O_2} + C_{O_2}} C_L C_{H^+} \quad (4)$$

where C_{O_2}, C_{H^+} are the concentrations of dissolved oxygen and protons in the volume of the electrolyte, $mol \cdot m^{-3}$; C_L is the surface concentration of the laccase enzyme at the cathode, $mol \cdot m^{-3}$; n_c – the number of electrons which take part in the electrode reaction; η_c is the local overpotential of the cathode, the difference between the electrode (cathode) and electrolyte potentials, V; k_c is the rate constant of the cathodic reaction, m·m$^3_{AL}$·mol^{-1}·s^{-1}; K_{O_2} – the half-velocity rate constant of dissolved oxygen, mol·m^{-3}; β – the charge transfer coefficient of the cathodic reaction.

Since a buffer solution was applied as the BFC electrolyte, it is assumed that the change in the concentration of protons is insignificant and it does not significantly affect the rate of electrode reactions. The surface concentration of laccase was determined in accordance with the procedure described in [8] and was also assumed to be constant.

According to [4] it was assumed that BFC can be treated as a continuously stirred tank reactor and all mass transfer processes are so fast compared to biochemical and redox reactions that the concentration of all components in the volume of the electrolyte can be considered to be equal in the whole bulk.

So, components concentrations changes vs time can be expressed by (5) - (8):

$$V_{lq}\frac{dC_G}{dt} = Q_G(C_G^0 - C_{O_2}) - A_a r_a \tag{5}$$

$$V_{lq}\frac{dC_{O_2}}{dt} = Q_{O_2}(C_{O_2}^0 - C_{O_2}) - A_c r_c \tag{6}$$

$$V_{lq}\frac{dC_{H+}}{dt} = 2A_a r_a - 4A_c r_c \tag{7}$$

$$V_{lq}\frac{dC_{GA}}{dt} = A_a r_a \tag{8}$$

where $C_{O_2}^0, C_G^0$ – the concentration of oxygen or glucose in the feed flow, mol·m^{-3}; V_{lq} – the volume of the electrolyte chamber, m^3; Q_{O_2}, Q_G – the feed flow rate of oxygen or glucose, m^3·s; A – a geometrical area of electrodes, m^2; C_{GA} is the gluconic acid concentration, mol·m^{-3}; indexes «*a*» and «*c*» denote the anode and cathode respectively.

The charge balances at the anode and cathode are given by eq. (9)-(10):

$$C_a\frac{d\eta_a}{dt} = i_{cell} - 2Fr_a \tag{9}$$

$$C_c\frac{d\eta_c}{dt} = -i_{cell} + 4Fr_c \tag{10}$$

where i_{cell} is the current density of the cell, A·m^{-2}; C_a and C_c are the capacitances of the anode and cathode, Coulomb·V^{-1}.

Similar to [4] it is assumed that the ohmic drops in the current-collectors and electric connections are negligible, and the cell resistance is solely due to resistance of the solution. So, the cell voltage is calculated as:

$$U_{cell} = U^0 - \eta_a + \eta_c - \frac{d_{cell}}{k^{lq}} i_{cell} \tag{11}$$

where: U^0 – is the open-circuit voltage, V; d_{cell} – the distance between electrodes, m; k^{lq} – the electrical conductivity of the solution, Ohm^{-1}·m^{-1}.

3. Results and discussion

3.1. Experimental results

The electrochemical characteristics of membraneless mediatorless BFC with electrodes' catalysts based on laccase and gold-modified carbon black in a phosphate-acetate buffer solution are shown in Figure 2(a).

It can be assumed that in the direct current mode, more efficient use of the anode active layer is achieved by increasing the number of reacting glucose molecules. In the case of a laccase-based cathode, the dependence of the BFC characteristics on the measurement mode is less apparent, which may be due to the limitation of the current generation mainly by the cathode process, which in turn depends on the amount of laccase involved in the oxygen reduction reaction (ORR). The corresponding data are presented in Figure 2(b).

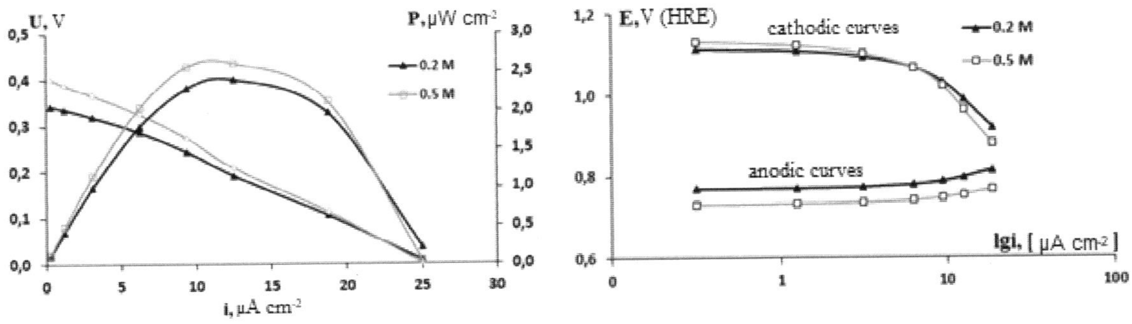

(a) (b)

Figure 2. Electrochemical characteristics of a BFC with a laccase-based cathode and a 20Au/XC-72R-based anode in a phosphate-acetate buffer solution (pH 4.7) containing 0.2 and 0.5 M glucose: a) Stationary volt-ampere curves and power density versus current density, b) Partial cathode and anode curves.

As can be seen, the ORR overpotential at the cathode with immobilized laccase when measuring the stationary curve is about 400 mV versus 80 mV for the Au20/XC-72R-based anode. This causes a slight effect of glucose concentration on the total volt-ampere curve (Figure 2 (a)), despite a significant (more than one order) increase in the speed of the anode process (Figure 2 (b)). On the other hand, the absence of a noticeable effect of glucose concentration on the speed of the cathode process indicates a high tolerance of laccase to the fuel used.

To assess the practical significance of the work, lasting tests of BFC in the galvanostatic mode were carried out. It was found that the stable functioning of the BFC with an enzyme is possible only when using extremely low current densities (1 $\mu A \cdot cm^{-2}$), while during tests for more than 150 hours, the voltage of BFC with a laccase-based cathode and an anode based on gold-modified carbon black fell from 0.32 to 0.23 V.

Besides, a significant decrease in the rate of oxygen recovery was observed during the transition from the electrolyte with pH 4.7 to pH 8, while the maximum power density of the entire BFC increased from 2.6 $\mu W \cdot cm^{-2}$ (pH 4.7) to 44 $\mu W \cdot cm^{-2}$ (pH 8), probably due to an increase in the activity of the anode. This aspect was not the subject of the study and will be studied in future works.

3.2. Results of mathematical modeling

Calculations were carried out for two cases of glucose concentrations: 0.2 and 0.5 M. The system of ordinary differential equations (1-11) was solved using numerical methods in MATLAB. 2011b.The operating conditions, experimental set-up parameters, and literature based constants are listed in Table 1. Parameters listed in Table 2 were fitted to match calculated values to the experimental one.

Table 1. Parameters of experimental set-up and operation, and constants of the BFC model.

Parameter	Description	Value
F, C mol^{-1}	Faraday's constant	96485.4
R, J mol^{-1} K^{-1}	Gas constant	8.3144
T, K	Temperature	303
d_{cell}, m	Distance between anode and cathode in the cell	$5 \cdot 10^{-3}$
C_a, F m^{-2}	Capacitance of anode [4]	400
C_c, F m^{-2}	Capacitance of cathode [4]	500
V_{lq}, m^3	Volume of electrolyte chamber	$7 \cdot 10^{-6}$
Q_{O_2}, m^3 s^{-1}	Flow rate of oxygen to cathode	$13.89 \cdot 10^{-8}$
Q_G, m^3 s^{-1}	Flow rate of glucose to cathode	0
$C^0_{O_2}$, mol m^{-3}	Oxygen concentration in the feed flow	0.2575

IOP Conf. Series: Earth and Environmental Science **633** (2021) 012001 doi:10.1088/1755-1315/633/1/012001

C_G^0, mol m^{-3}	Glucose concentration in the feed flow	0
C_L, mol m$^{-3}_{AL}$	Laccase concentration in the active layer	$0.468 \cdot 10^{-3}$
A_c, m^2	Area of cathode	$16 \cdot 10^{-4}$
A_a, m^2	Area of anode	$16 \cdot 10^{-4}$
U^0, V	Cell open circuit potential for:	
	200 mol/m^{3}glucose concentration	0.344
	500 mol/m^{3}glucose concentration	0.4
K_{O_2}, mol m^{-3}	Half velocity rate constant for dissolved oxygen[4]	0.004
C_{O_2}(t=0), mol m^{-3}	Initial oxygen concentration	0.2575
C_G (t=0), mol m^{-3}	Initial glucose concentration	200, 500
C_{H+}(t=0), mol m^{-3}	Initial protons concentration of the electrolyte, based on pH	0.02

Table 2. Estimated parameters of glucose BFC.

Parameter	Description	Value
k_c, m·m$^3_{AL}$ ·s^{-1} ·mol^{-1}	Forward rate constant of cathode reaction	$1 \cdot 10^{-5}$
k_a, m·s^{-1}	Forward rate constant of anode reaction	$1 \cdot 10^{-8}$
α, dimensionless	Charge transfer coefficient of anode	0.6
β, dimensionless	Charge transfer coefficient of cathode	0.2
k^{lq}, Ohm^{-1} m^{-1}	Electrical conductivity of the electrolytefor :	
	200 mol/m^{3}glucose concentration	0.004
	500 mol/m^{3}glucose concentration	0.0032

The results of calculations based on mathematical modeling are shown in figure 3-6. Figure 3 shows experimental and calculated discharge characteristics for BFC with laccase based cathode and 20Au/XC-72R based anode in phosphate acetated buffer solution (pH 4.7), with glucose 0.2 M and 0.5 M concentration. As it is shown in figure 3, there is a good agreement between the calculated and the experimental data. The deviation was calculated according to (12) and was 0.0007 for experiments with 0.2M glucose concentration and 0.0196 for the case with 0.5M glucose concentrations:

$$\Phi = \sqrt{\frac{1}{N}\sum_{j=1}^{N}\left(i_j^{\exp} - i_j^{\text{calc}}\right)^2} \tag{12}$$

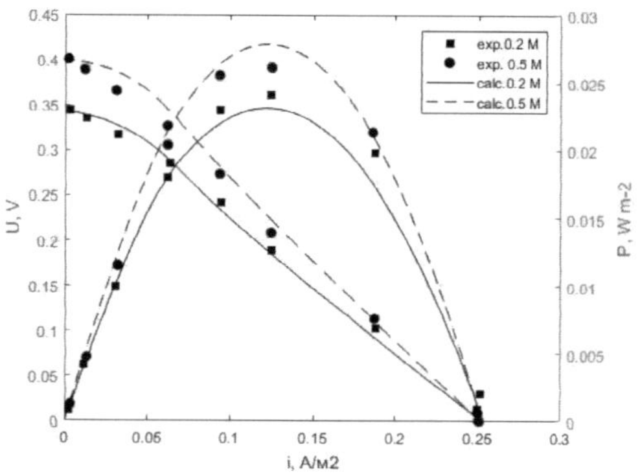

Figure 3. Experimental and calculated discharge characteristic for BFC with laccase based cathode and 20Au/XC-72R based anode in phosphate acetated buffer solution (pH 4.7), with glucose 0.2 M and 0.5 M concentration.

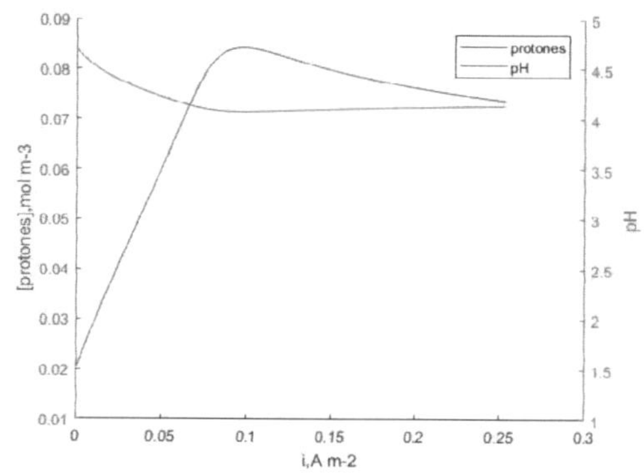

Figure 4. Dependence of protons concentration in the bulk of the electrolyte vs current density of BFC (for 0.2 M glucose concentration).

Figure 4 shows the change of pH and protons concentration in the electrolyte solution during potential-dynamic experiments. One can see in figure 4 that pH slightly changes (decreases from 4.7 to 4.1) as the result of protons production in glucose oxidation reaction on the anode. At current density equals to 0.1 $A \cdot m^{-2}$ one can see protons consumption because of oxygen reduction reaction (ORR). It should be mentioned that the model does not take into account the nature and behaviour of the buffer solution.

Figure 5 shows the dependencies of glucose and oxygen concentrations vs current density as the time of potential-dynamic experiments. It should be mentioned that the glucose concentration in the solution slightly changes. The time of potential-dynamic experiments is not enough to make a conclusion about the character of the concentration changes. But observing the tendency to decreasing shows fuel (glucose) consumption during the oxidation reaction.

Profile of oxygen concentration shows slight concentration drop for oxygen consumption at ORR, but the changes are not significant for oxygen continuous flow into the system.

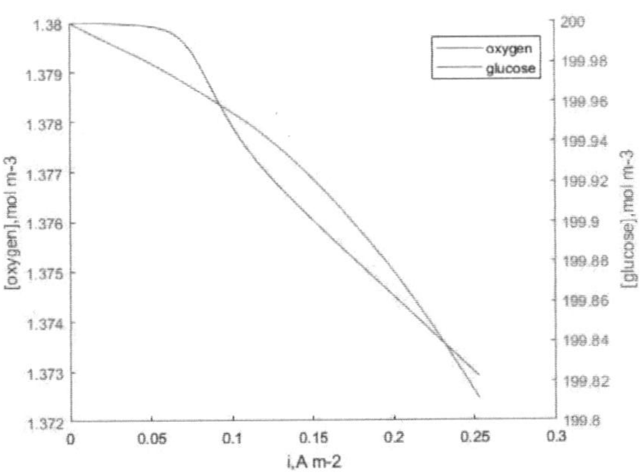

Figure 5. Dependencies of glucose and oxygen concentrations in the bulk of the electrolyte vs current density of BFC (for 0.2 M glucose concentration).

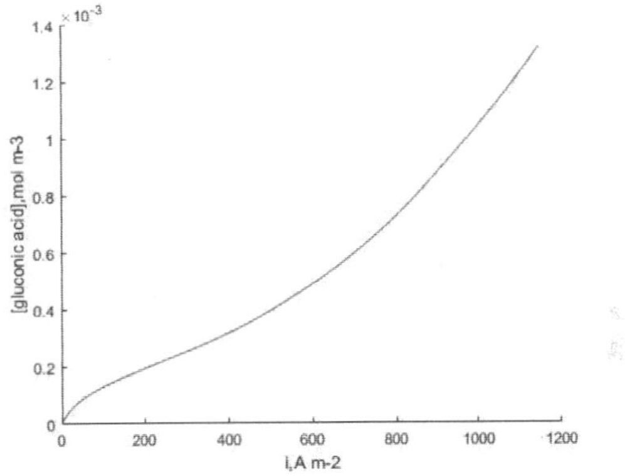

Figure 6. Dependence of gluconic acid concentration in the bulk of the electrolyte vs current density of BFC (for 0.2 M glucose concentration).

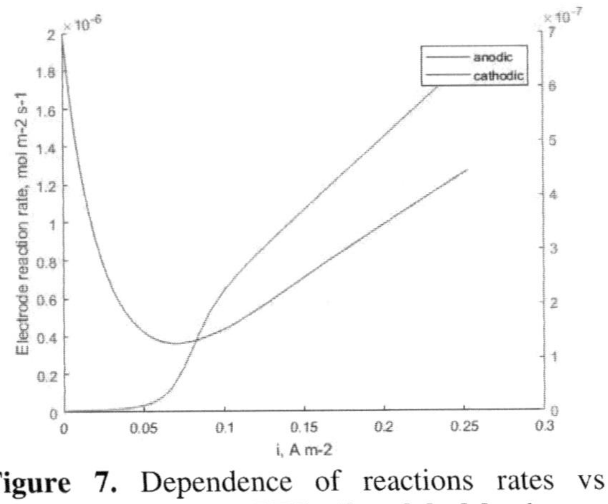

Figure 7. Dependence of reactions rates vs current density of BFC (for 0.2 M glucose concentration).

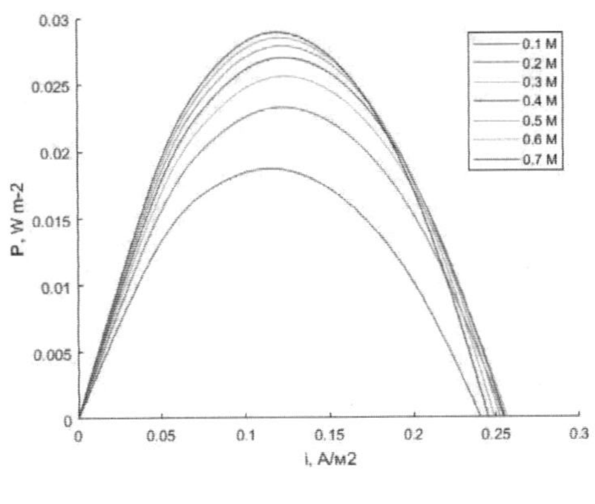

Figure 8. Calculated values of power density vs current density of BFC.

From figure 5 one can see that the glucose concentration decreases with the current density increasing in the cell, meanwhile, the concentration of the gluconic acid rises (figure 6). It can be explained by the fact the higher value of the current density leads to the faster rate of the reaction, in other words to quicker fuel consumption and gluconic acid production.

It was determined that with the rise of the current density the local overpotential of anode increases while the local overpotential of cathode decreases. These results correlate with decreasing of the cell voltage.
Profiles of anodic and cathodic reaction rates vs time are shown on figure 7. The rate of cathodic reaction is significantly lower than anodic one and the cathodic process is limiting.
A search for unknown values of the electrical conductivity of the electrolyte solution and the open-circuit voltage for each concentration was found on the basis of logarithmic dependencies based on the data for known glucose concentrations of 0.2 and 0.5 M. Referent curves are the following:

$$k_{lq} = -9 \cdot 10^{-4} \cdot \ln\left(C_G^0\right) + 0.0088 \quad (13)$$

$$\mathrm{U}^0 = 0.0611 \cdot \ln\left(C_G^0\right) + 0.0202 \quad (14)$$

Figure 8 shows the effect of glucose concentration on the cell power output. An increase in glucose concentration causes cell power growth.
Since the fuel concentration is considered to be an important factor affecting the performance of BFCs, series of calculations with different values of fuel (glucose) concentration in the range of 0.1-1.5 M were carried out in order to optimize the operating modes.
Optimization was performed by the Nelder-Mead method. The optimum value of the concentration of the supplied fuel, which provides the maximum value of the power density at the highest value of the current density is 0.546 M.

4. Conclusion

In this paper, experimental studies of a membraneless mediatorless glucose-oxygen BFC with catalysts based on laccase and gold-modified carbon black were carried out. It was shown that in a slightly acidic medium (pH of the buffer solution was 4.7) and 0.5 M of glucose concentration, it achieved the maximum power density of 2.6 μW/cm^2.
A mathematical model that allow us to explore the processes of the studied BFC in detail and can help to reduce the load on laboratory research was developed. As a result of calculation by the developed mathematical model, the concentration distribution of the main participants of electrode reactions on the current density, discharge curves, dependences of BFC's power density on current density, overpotentials of the electrodes, the dependences of electrodic reactions rates on the current density of BFC and other data were obtained for 0.1-1.5 M range of glucose concentration in a nutrient solution. As a result of the optimization, it was found that the optimum of the glucose-containing is 0.546 M. Besides, it was found that the electrical properties of the buffer solution have some influence on the output electrochemical characteristics of the device. In the future, we plan to expand the use of the developed mathematical model on a BFC which is operating in slightly alkaline conditions, since, as it was found, this provides increasing of the output characteristics of the device.

5. References

[1] Bogdanovskaya V A, Arkad'eva I N, Osina M A 2017 Russ. J. Electrochem. 53 (12) 1323
[2] Gorshina E S, Rusinova T V, Biryukov V V, Morozova O V, Shleev S V, Yaropolov A I 2006 Appl.Biochem.Microbiol. 42 558
[3] Tripachev O V, Tarasevich M R, Maleeva E A 2011 Electrochemical Energetics 11 (1) 11
[4] Zeng Y, Choo Y F, Kim B H, Wu P 2010 J. Power Sources 195 79
[5] Arkadeva I, Fokina E, Bogdanovskaya V, Vasilenko V, Koltsova E 2018 GeoConferences SGEM 441.
[6] Vasilenko V A, Arkadeva I N, Fokina E A, Koltsova E M , Bogdanovskaya V A 2018 CET Journal 70 1609
[7] Megias-Sayago C, Bobadilla L F, Ivanova S, Penkova A, Centeno M A, Odriozola J A 2018 Catal. Today 301 72
[8] Arkadeva I N, Bogdanovskaya V A, Vasilenko V A, Fokina E A, Koltsova E M 2017 IOP Conf. Ser.: Earth and Environ. Sci. 83 127

IOP Conf. Series: Earth and Environmental Science **633** (2021) 012002 doi:10.1088/1755-1315/633/1/012002

Design, fabrication, and testing of biomass pelleting machine for coffee wastes

G J C Bartolome[1], S M C De Leon[2], C A Polinga[2] and J M B Roño[3]

1 Department of Engineering, [3]Research and Extension Unit, Cavite State University-Cavite College of Arts and Trades, Tejeros Convention, Rosario, Cavite, Philippines
2 Department of Agricultural and Food Engineering, Cavite State University-Main Campus, Indang, Cavite, Philippines
E-mail: geejaybartolome@gmail.com

Abstract A biomass pelletizer was designed and fabricated to produce pellets using coffee husks and spent coffee grounds. The pelletizer works on a principle of a rotating roller pellet mill in which power from the prime mover is transmitted to the rotating spindle at a reduced speed, and causes the roller dies to drift and press the pellet mixture through the thick perforated die plate. An unsharpened knife attached to the spindle facilitates easy detachment of pellets from the die plate upon extrusion and the perforated pellet chute separates the non-pelletized particles and fines from the pellets itself. The performance evaluation of the machine revealed an average pelleting capacity and efficiency of 1.50 kgh^{-1} and 92.49 percent, respectively, suitable for small-scale pellet production and also a low percentage losses (5%). The physical characteristics of the pellets produced were found to have high uniformity with an average length of 9.86±0.63 mm, diameter of 5.71±0.41 mm, mass per pellet of 0.12 g, moisture content (dry basis) of 16.65 percent, and a bulk density of 850 kg m^{-3}. The researchers recommend that the performance of the pelleting machine be optimized by attaching a feed controller on the hopper and manipulating the speed of the roller dies. Furthermore, it will also be best to analyze the thermochemical properties of the pellets including the proximate and ultimate composition employing standard test methods.

1. Introduction

There is a growing interest on the use of biomass for heat and energy generation with the need to reduce fossil fuel consumption, greenhouse gas emissions, and environmental degradation [1-2]. Among renewable energy alternatives, biomass has the major advantage of being available in large quantities from a variety of sources, and they can be utilized for sustainable heat and power production with significant environmental and socioeconomic benefits. Large volume of biomass comes from agricultural wastes, however, they lack the required density to be transported to the point of energy production causing expensive transportation cost [3]. Moreover, biomass in raw form, has lower volumetric energy content than traditional fossil fuel, low bulk density, high moisture content and hydrophobic character [2, 4, 5]. Agricultural residues and woody biomass have bulk densities ranging from 80-100 kgm^{-3} and 150-200 kgm^{-3}, respectively, that can be densified up to 700 kgm^{-3} through pelletization [6]. Through this process, biomass materials are provided with uniform physical properties and improved combustion properties at low moisture content, and costs of handling, transportation, and storage are significantly cut down.

In general, a large percentage by weight of biomass feedstocks consist of cellulose (40-50%), followed by hemicellulose (20-40%), and lignin. Along with the relative percentage composition of biomass, key factors that determine the suitability as energy crops include the amount of moisture and ash, fixed carbon, calorific value, and alkali metal content [3, 7]. The main raw materials used in pelletization are wood residues like saw dust, wood shavings and chips, as well as agricultural residues, food industry wastes and fuel crops in which a pellet mill with ring roller dies are commonly employed in commercial facilities due to their high capacity and good pelletization properties [8-9]. The configuration and operation of the pellet mills greatly influence the efficiency of the operation and properties of the pellets produced in addition to the type of feedstock, additive or binders, and other machine parameters [10-11]. To maintain homogenous pellet products in the national and international markets, various quality standards have been prescribed [12]. Biomass pellets also generate lower emissions over their original form and other conventional biomass fuels [13-14]. In addition, its dimensional configuration and standard size allow a more compact storage, handling, and automatic feeding in large scale operations.

Depending on the kind of processing, coffee generates a significant amount of agro-waste. Coffee pulp and coffee husks are among the solid residues and currently have no profitable uses [15]. Coffee husks which is fibrous in nature, consist of organic-rich components (50% cellulose, 38% hemicelluloses, pectin, and lignin) that makes it an ideal feedstock for pellet production. Its composition is also similar to wood dust. Meanwhile, spent coffee grounds (SCG), is another major by-product of coffee processing and it was estimated that about 2 kg of SCG is produced per kilogram of instant coffee, generating several millions of tons of waste worldwide every year [16]. SCG is a solid residue that has high amount of moisture, acidic, and organic-rich, yet, its composition is very diverse, suggesting that it can be used for a variety of applications. SCG is composed of a huge amount of carbon-based compounds like the coffee husks, wherein, cellulose is the primary cause of volatile matter and lignin plays as a natural binder [17].

Studies on biomass pelletizing have been carried in different ranges from small to large-scale operations using various configurations of pelleting machines. On the other hand, biomass feedstocks are commonly blended to improve the properties of the pellets, but few studies have focused on using coffee wastes and pure biomass materials. There is also a limited availability of local pelleting machines that can be used in small-medium scale operations. Adoption of machines from foreign manufacturers may also pose problems or difficulties in maintenance and after-sales services. To resolve these issues, a locally-fabricated pellet machine suitable for small-medium scale operations was designed employing the principle of roller pellet mills. Various mixes of coffee husks and SCG was also used as pellet feedstock. Preliminary investigations focused on the efficiency and capacity of the pelleting machine and the uniformity of the properties of the pellets. It is deemed that this project will open up new applications of the coffee wastes while coming up with a locally-made, easy to operate, and economically-feasible agricultural machinery.

2. Materials and methods

2.1. Design criteria

The pelleting machine should convert the pellet mixture of coffee husks and spent coffee grounds into densified pellets. It follows the principle of a rotating roller pellet mill where power is transmitted from the electric motor to a shaft or spindle with a reduced speed, and urges the roller dies to drift and press the pellet mixture. The machine was designed such that at least one person is required to control its operation, materials for construction were low-cost and locally available, and the fabrication procedures should include basic machining operations such as drilling, boring, and milling for ease of machine manufacturing.

2.2. Design details and specifications

The pelleting machine (Figure 1) is a motor-operated machine powered by a single-phase 1.5-hp electric motor. The speed of the motor was reduced to approximately 30 rpm from 1800 rpm using a 1:60 speed reducer. It has an overall dimensions of 740 x 413 x 790 mm (*L x W x H*). High-tensile

IOP Conf. Series: Earth and Environmental Science **633** (2021) 012002 doi:10.1088/1755-1315/633/1/012002

fasteners were used for assembly of the components to reduce risks of misalignment and breakage due to high pressure and temperature build-up during the pelleting operation.

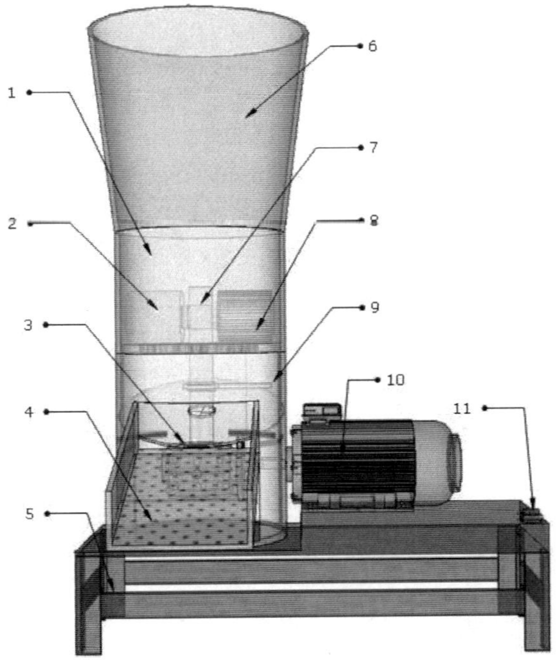

Legend:

1. pelleting chamber
2. flat roller die
3. gearbox
4. perforated pellet chute
5. main frame
6. hopper
7. roller shaft
8. grooved roller die
9. knife
10. prime mover
11. switch

Figure 1. Representation of the biomass pelleting machine

The machine consists of several major components that allow a highly efficient pelleting operation. The pellet mixture is loaded on a frustum-shaped hopper with dimensions of 215 x 245 mm (*H x dia.)* that directs the mixture to the pelleting chamber, with a holding capacity of 1 kg. The pelleting chamber facilitates mixing and pressing of the pellet mixture using two roller dies – one grooved type and one flat type. The grooved type roller die has 24 teeth and each tooth is 2 mm x 5.67 mm *(H x W)*. The roller dies were made of cold-rolled steel 1080 and were attached to the shaft perpendicularly, allowed to freely rotate on its axis through deep-groove ball bearings. The pressed mixture is extruded into cylindrical-shaped solid pellets with approximately 6-mm diameter through the thick die plate, which is locked in position by bolts and nuts to prevent slippage during the operation. The thick die plate has 96 holes whose inlet and outlet diameters of each hole were 10 mm and 6 mm, respectively. The high pressure build-up inside the chamber increases its frictional forces and so its temperature. A knife attached to the shaft cut the produced pellets from the die plate to a desired length of 10 mm. Upon extrusion, pellets are discharged to an outlet chute, inclined at 30°, with 4-mm perforations to separate the non-pelleted particles and fines from pellets.

2.3. Testing and evaluation

Pelleting was carried out in three trials using 1 kg of pellet mixture in each trial. The mixture was composed of the following: 28.6 percent coffee husks, 28.6 percent spent coffee grounds, 14.2 percent starch, and 28.6 percent water. The performance characteristics of the components of the machine including the speed of prime mover shaft, spindle, and roller shaft, the noise level, and the input power of the machine were determined. The machine was also evaluated in terms of its pelleting capacity and efficiency with the procedure adopted from the Philippine Agricultural Engineering Standards for hammer mill (PAES 217:2004).

The pelleting capacity refers to the weight of pellets produced per unit of operating time (kgh^{-1}) while pelleting efficiency is the ratio of the weight of pellets produced on the pellet discharge chute to the total weight of biomass materials fed into the hopper (%), computed as follows:

$$\text{pelleting capacity,} \quad C_p = \frac{W_p}{T_p} \tag{1}$$

$$\text{pelleting efficiency,} \quad Ep = \frac{Wp}{Wi} \tag{2}$$

where C_p is the pelleting capacity (kgh^{-1}), Wp is the weight of pellets produced (kg), T_p is the total time of pelleting (h), E_p is the pelleting efficiency, and W_i is weight of input biomass mixture (kg).

2.4. Physical properties of pellets

The physical properties of the pellets were described in terms of pellet length, pellet diameter, pellet mass and bulk density. The dimensions of each of the 100 randomly selected pellet samples from each trial were measured.

In determining the moisture content, three 50-gram samples of pellets from each trial were oven-dried at 110°C for 24 hours or until constant weight. The moisture content of each sample was calculated using the equations below:

$$\% \, MC_{wet} \, \text{basis} = \frac{(Mo - M_1)}{Mo} \times 100\% \tag{3}$$

$$\% \, MC_{dry} \, \text{basis} = \frac{(Mo - M_1)}{M_1} \times 100\% \tag{4}$$

where $\%MC$ is the moisture content (wet basis and dry basis), M_o is the initial mass of the pellet samples (g) and M_1 is the final mass of the oven dried pellet samples.

For the determination of the bulk density, cooled pellets were loaded into a funnel placed in a tared container with known volume. Pellets were allowed to free flow into the container. Excess pellets (above the top rim of the container) were removed by scraping off with a straight edge. The container and pellets were weighed and the weight per volume was reported as its bulk density.

3. Results

3.1. Preliminary testing observations and interventions

The performance of the pelleting machine varied with different mixtures of coffee husks, spent coffee grounds, binder, and water. Among the problems encountered were longer pelleting time primarily due to the congestion in the holes of the flat die plate, stacking up and hardening of the mixture, and occurrence of occasional stops when the spindle and roller dies operate. To overcome these problems, the large particle size of the coffee husks in the mixture were reduced to ease passage of pellets through the die plate. The amount of binder was also optimized to improve the firmness of the pellets and avoid cracking. This also helped reduce the chances of hardening of the mixture in the pelleting chamber. Lastly, to avoid congestion in the die plate, feeding of the pellet mixture was manually controlled.

3.2. Performance characteristics

The performance of the pelleting machine was characterized with and without load wherein, no significant difference was observed among parameters (Table 1). The average speed of the motor shaft was recorded at 1,799.0 rpm and the speed of the roller dies was 28.1 rpm. Noise levels were below the limit for agricultural machineries 26.5 db (<92 dB).

The evaluation of the pelleting capacity and pelleting efficiency revealed an average value of 1.50 kgh^{-1} and 92.49 percent, respectively. At this rate, the percentage losses reached only 7.5 percent of the total input mixture while 5.0 percent of mixture is lost during the pelleting operation. When compared with other pelleting machines, this average pelleting capacity of may be low primarily due to lower roller shaft speed that affects the rate of pelleting. Nonetheless, the recorded pelleting efficiency is a good indicator to the performance of the machine.

3.3. Physical properties of pellets and pellet uniformity

The average length and diameter of the pellets produced were 9.86±0.63 mm and 5.71±0.41 mm, respectively (Table 2). The machine was designed to produce pellets with average length of 10 mm and average diameter of 6 mm. Statistical analysis showed that the 75th percentile of the 300 pellet samples has length ≥10.39 mm and diameter of ≥5.82 mm. The average mass of each pellet was also almost uniform at 0.12±0.001 g.

In terms of the moisture content and the bulk density, the average moisture content in dry and wet basis of the pellets is 16.65 and 14.27 percent, while the bulk density is 850 kgm^{-3}. Similar raw materials (spent coffee grounds) upon pelleting were also reported by earlier researchers. A density of 1058-1255 kgm^{-3} and 6.2-11.78% for moisture content were reported for pellet's characteristics of spent coffee grounds [17-19]. However International Organization for Standardization standardized the fuel specifications for solid biofuels, a density higher than 500 kgm^{-3} and a moisture content lower than 10% for pellet must be achieved for its safe storage and efficient combustion [20]. This is supported by other research works stating that a density of 700 kgm^{-3} and a moisture content of 8-12% for pellets is a typical bulk densities and moisture contents of biomass pellets and a pellet's density below 400 kgm^{-3} is a low density pellet for biomass making it difficult to store, transport and utilize [21-22].

Table 1. Observations on the pelleting efficiency and pelleting capacity of the pelleting machine.

Performance Characteristic	Mean[a]
Speed of the motor shaft	1799.0 rpm
Speed of the roller die(s)	28.1 rpm
Noise level	265. dB
Power input	2.0 kW
Current drawn	8.4 A
Voltage supply	238.3 V
Total weight of input	1000.8 g
Weight of pellets produced	925.7 g
Weight of fines	24.6 grams
Losses	50.5 grams
Percentage Losses	5.0%
Pelleting time	0.6 hr
Pelleting Capacity	1.5 kg h^{-1}
Efficiency	92.49%

[a]*Difference in the mean values is not significant.*

Table 2. Analysis of the physical properties of 300 pellet samples in terms of length, diameter, and mass for uniformity assessment

Parameter		Physical Property		
		Length	Diameter	Mass
N	Valid	300	300	300
	Missing	0	0	0
Mean		9.8609	5.7053	.1160
Variance		.631	.0410	.001
Minimum		5.06	4.7600	.10
Maximum		12.45	6.1700	.20
Percentiles	25	9.3800	5.6100	.1000
	50	9.9000	5.7350	.1000

75	10.3850	5.8200	.1000

4. Conclusion
The developed biomass pelleting machine can produce pellets from a mixture of coffee husks and spent coffee grounds at an acceptable pelleting capacity and high efficiency. The pellets produced has uniform physical characteristics and low moisture content that reflect the reliability of the machine. Suitable for small-scale pelleting operation, the machine can be easily manufactured using locally available materials. The pelleting capacity can be increased by optimizing the speed of the spindle and roller dies, coupled with controlled feeding. It is recommended for further studies to include the analysis of the calorific value of the pellets and their proximate and ultimate composition using standard methods. Furthermore, to assure that the pellets will be safe for long-term storage, air-drying is recommended to reduce its moisture content after processing.

5. References
[1] Iftikhar M, Asghar A, Ramzan N, Sajjadi B and Chen W 2019 Biomass densification: Effect of cow dung on the physicochemical properties of wheat straw and rice husk based biomass pellets *Biomass Bioenergy* **122** 1-16
[2] Stelte W, Holm J K, Sanadi A R, Barsberg S and Ahrenfeldt J 2011 Fuel pellets from biomass: The importance of the pelletizing pressure and its dependency on the processing conditions *Fuel* **90** 3285-90
[3] Bajwa, D S, Peterson T, Sharma N, Shojaeiarani J and Bajwa S G 2018 A review of densified solid biomass for energy production *Renew. and Sust. Energ. Rev.* **96** 296-305
[4] Jackson J, Turner A, Mark T and Montross M 2016 Densification of biomass using a pilot scale flat ring roller pellet mill *Fuel Processing Technology* **148**, 43-49
[5] Jamradioedluk J and Lertsatitthanakorn C 2017 Influences of mixing ratios and binder types on properties of biomass pellets *Energy Procedia* **138** 1147-52
[6] Soleimani M, Tabil X, Grewal R and Tabil L Carbohydrates as binders in biomass densification for biochemical and thermochemical processes *Fuel* **193** 134-41
[7] McKendry P 2002 Energy production from biomass (part 1): overview of biomass *Bioresour. Technol.* **83** 37-46
[8] Sultana A, Kumar A and Harfield D 2010 Development of agri-pellet production cost and optimum size *Bioresour. Technol.* **101** 5609-21
[9] Larsson S H and Rudolfsson M 2012 Temperature control in energy grass pellet production – effects on process stability and pellet quality *Appl. Energy* **97** 24-9
[10] Harun N Y, and Afzal M T 2016 Effect of particle size on mechanical properties of pellets made from biomass blends *Procedia Eng.* **148** 93-9
[11] Nielsen N P K, Gardner D, Poulsen T and Fely C 2009 Importance of temperature, moisture content, and species for the conversion process of wood residues into fuel pellets *Wood Fiber Sci.* **41(4)** 414-25
[12] Pradhan P, Mahajani S M and Arora A 2018 Production and utilization of fuel pellets from biomass: A review *Fuel Process. Technol.* **181** 215-32
[13] Ghafghazi S, Sowiati T, Sokhansanj S, Bi X and Melin S 2011 Particulate matter emissions from combustion of wood in district heating applications Renew. Sust. Energ. Rev. **15** 3019-28
[14] Shen G, Tao S, Wei S, Zhang Y, Wang R, Wang B, Li W, Shen H, Huang Y, Chen Y, Chen H, Yang Y, Wang W, Wei W, Wang X W, Liu W, Wang X and Simonich S L M 2012 Reductions in emissions of carbonaceous particulate matter and polycyclic aromatic hydrocarbons from combustion of biomass pellets in comparison with raw fuel burning *Environ. Sci. Technol.* **46** 6409-16
[15] Oliveira L S, and Franca A S 2015 Chapter 31 – An overview of the potential uses for coffee husks *Coffee in Health and Disease Prevention* 283-291
[16] Martinez-Saez M, Garcia A T, Perez I D, Rebollo-Hernanz M, Mesias M, Morales F J, Martin-Cabrejas M A and Del Castillo M D 2017 Use spent coffee grounds as food ingredient in bakery products *Food Chem.* **216** 114-22

[17] Lisowski A, Olendzki D, Swietochowski A, Dabrowska M, Mieszkalski L, Ostrowska-Ligeza E, Stasiak M, Klonowski J and Piatek M 2019 Spent coffee grounds compaction process: its effects on the strength properties of biofuel pellets *Renew. Energy* **142** 173-83

[18] Jeguirim M, Limousy L and Fossard E 2016 Characterization of coffee residues pellets and their performance in a residential combustor *International journal of green energy* **13**(6) 608-15

[19] Limousy L, Jeguirim M, Dutournie P, Kraiem N, Lajili M and Said R 2013 Gaseous products and particulate matter emissions of biomass residential boiler fired with spent coffee ground pellets *Fuel* **107** 323-29

[20] ISO 17225-6 Solid Biofuels Fuel Specifications and Classes Part 6: Graded Non-woody Pellets English Version EN 2014

[21] Sokhansanj S, Turhollow A F 2004 Biomass densification – cubing operations and costs for corn stover *Appl. Eng. Agric.* **20** 495-9

[22] Tumuluru J S, Wright C T, Kenney K L and Hess J R 2010 *A technical review on biomass processing: Densification, Preprocessing, Modeling and Optimization* (Pittsburg, Pennsylvania: ASABE)

IOP Conf. Series: Earth and Environmental Science **633** (2021) 012003 doi:10.1088/1755-1315/633/1/012003

Comparison of actual Levelized Cost of Electricity of solar thermal concentrating solar power plants

A Boretti[1], S Castelletto[2], W Al-Kouz[1,] and J Nayfeh[1]
1 Mechanical Engineering Department, College of Engineering, Prince Mohammad Bin Fahd University, Al Khobar, Saudi Arabia
2 School of Engineering, Royal Melbourne Institute of Technology (RMIT) University, Bundoora, Victoria, Australia
E-mail: a.a.boretti@gmail.com

Abstract. Forecasts based on modeling of concentrated solar tower power plants with thermal energy storage of 10 hours daily predict the production of dispatchable electricity at 6 c/kWh. Currently, there is no plant utility size in operation featuring this technology and at this predicted cost. The recent experience of Crescent Dunes, an example of this type of technology, should be used to better estimate the current cost of dispatchable electricity. Crescent Dunes started operation in October 2015 demonstrating since the very beginning the lack of maturity of this specific technology, experiencing a lack of production or no production at all, every single month since starting operation. The 110 MW plant (currently not in operation) provides a cost, excluding repairs and maintenance costs, of 2.38 $/kWh of unpredictable electricity. This experience suggests that every prediction of costs and performances should be based on data of plants built and operating, to avoid the use of models not yet validated to predict performances of novel plants. By surveying other technology based on parabolic through, the current actual costs are here determined for specific latest operating plants in the range of more realistic 8-12 c/kWh.

1. Introduction

Most of the predictions of energy production and the Levelized Cost of Electricity (LCOE) of large solar power plants are based on models and little empirical data from smaller plant validation. In the specific, NREL [1] forecasts as the best technology for Concentrated Solar Power the Solar Tower technology with 10 hours of two tanks thermal energy storage by molten salt. As an example, the facility Crescent Dunes was one of these examples and the first large molten-salt power tower plant in the United States. It was commissioned in 2015 with a reported installed CAPEX of 8.96 c US$/kW. Crescent Dunes was also the first and only plant utility size of this kind built in the world and never operated as expected during the lifetime, with many months without production, with a delivered capacity factor of 13% compared to an expected capacity factor of 52% based on a forecast.

NREL predicts that this specific technology should cost in 2017 7,800 US$/kW, Fig.1. a. Crescent Dunes had, however, a cost in 2015 of 0.975 billion US$, corresponding to 8,864 US$/kW. Furthermore, NREL was forecasting an even larger reduction of capital costs since 2017 with the CAPEX of 2020 further down to 6,500 US$/kW. This estimation seems affected by large inaccuracies when capacity factors are considered and the Levelized Cost of electricity is determined. CAPEX is the cost per installed capacity (nominal power). The capacity factor is the ratio of actual generating power to nominal power. As the resource is not constant, capacity factors are very far from unity. NREL has also forecast that this specific technology of Concentrated Solar Power the Solar Tower technology with 10 hours of two tanks thermal energy storage by molten salt should have delivered in 2017 capacity factors from 50% to 64%,

Fig.1.b. This forecast is attributed to a range of solar resources such as Abilene, Texas, with about 6.16 PSH/day, Phoenix, Arizona, with 7.26 PSH/day and Daggett, California with 7.65 PSH/day. Tonopah, Nevada where Crescent Dunes is located is much closer to an excellent resource location than a good resource location, with about 7.51 PSH/day, and Crescent Dunes was working at about 13%.
According to ref. [1], Fig.1 a and b, this specific technology should cost today 30% less than the cost of Crescent Dunes, while it should produce 20% more than the electricity production predicted for the same, that is almost 400% more than the actual electricity delivered by the same. While referring to only one case to invalidate predictions can be incorrect, it appears that those predictions were based on no cases at all given the large inaccuracy.
As discussed in ref. [2] to [5], the much better performing, simpler and more established parabolic trough technology, without thermal energy storage can achieve actual capacity factors approaching 30% (Genesis, Mojave, 28.5% in the best year so far), spending less than Crescent Dunes, or with a smaller 6 hours' thermal energy storage capacity factors approaching 40% (Solana, 36.5% in the best year so far).
The one and only "world largest" concentrated solar power solar tower with thermal energy storage of Crescent Dunes got burst, [6], [7], [8]. This follows the cancellation of projects by the same developer for the same technology worldwide. This has not prompted yet NREL to revise their technology forecast for concentrated solar power that is simply wrong, underrating the much better performing parabolic trough that is not considered at all, and overrating what has been so far a failure technology in the real world.
Here, we show how Crescent Dunes performed every month in the real world and the comparison to other plants with a more mature technology based on parabolic troughs.
This analysis prompts the need to validate the systems model with real production data, as only a few examples in the literature are trying to address this challenge. This validation could avoid future economical failure and provide a better forecast in solar power plants' large investments towards more performing technology already available.

2. Method

We analyze the electricity production by Crescent Dunes, Solana, and Genesis by using the publicly available data supplied by the USA EIA [9]. As described before, the 110 MW Cresent Dune plant is the only solar tower consisting of a solar field of heliostats, a receiver atop of a central tall tower, the power block, and the optional thermal energy storage. The solar field is made up of the many heliostats. They are flat, sun-tracking mirrors that focus the solar DNI onto the receiver. Thermal energy is collected by the heat transfer fluid heated in the receiver and delivered to the power block. Tracking is performed on two axes. Thermal energy storage of 10 hours was also promised. The capital cost is estimated at 1 billion$. A comparison with SAM [10] of the predicted energy production is provided in ref. [11], showing the need for more empirical data to the physical model and further validation.
The design is much more complicated compared to the parabolic trough collectors associated with Solana and Genesis. Solana is a 250 MW solar power parabolic through the plant in Arizona, southwest of Phoenix, completed in 2013, with molten salt thermal energy storage of 6 hours. Its cost was 2 billion $. Genesis Solar plant is a concentrated solar power parabolic through located in the Mojave Desert with no energy storage. The solar power plant consists of two independent 125 MW net sections for a total of 250 MW for a cost of 1.25 billion $ commissioned in 2014. Both Solana and Genesis started production in late 2013.
Here we first determine the capacity factors of the plants, the ratio of the average generating power over a period T, to the nominal power P,

$$CF = \frac{E}{P \cdot T},$$

where E is the energy produced in the period T. The Levelized cost of electricity is here simply determined, due to lack of knowledge of future CF in the specific years, by

$$LCOE \cong \frac{CapEx \cdot P + O\&M \cdot N \cdot P}{\overline{CF} \cdot N \cdot T},$$

where $CapEx$ is the capital cost per unit installed power, $O\&M$ is the operation and maintenance annual costs, N expected life of the plant, and $\overline{CF}$ is the average annual capacity factor and T is the number of hours in 1 year.

3. Results

Fig.2 presents the actual performances of the above-described plants over the years of production, where an increasing CF is observed for Solana and Genesis over time, well above the performance of Crescent Dune.

The $LCOE$ is computed considering the $CapEx$ based on construction cost per unit power, while the $O\&M$ are determined based on 66 $/kW/yr [1] attributed to a solar tower which has the highest costs among the concentrated solar technologies, assumed here 50$/kW/yr for Solana due to the additional energy storage and 33$/kW/yr for Genesis, which has no energy storage. The $\overline{CF}$ over the current years of operation for Solana is found to be 36% and for Genesis 28%, providing $LCOE$ of~11.7 and 8.1 c/kWh, respectively. The lifetimes of the plants are assumed 25 and 30 years respectively, based on other older parabolic through technology-based plants such as SEGS, still in operation, and commissioned in 1984.

By taking for Crescent Dunes a predicted CF of 52% and a CAPEX equal to the actual construction cost of 975 million US$ divided by a capacity of 110 MW, or 8863 $/kW this could have translated into a production of 501,072 MWh per year that repeated over 25 years of the life of the plant, could have translated in a cost of 9.2 cUS$/kWh. As the Crescent Dunes plant has been shut down after less than 4 years of operation, having produced only 418,849 MWh, without considering the O&M costs, and the repairs incurred to rectify damages to the plant, the US taxpayers thus paid unpredictable electricity from the sun 2.38 US$/kWh.

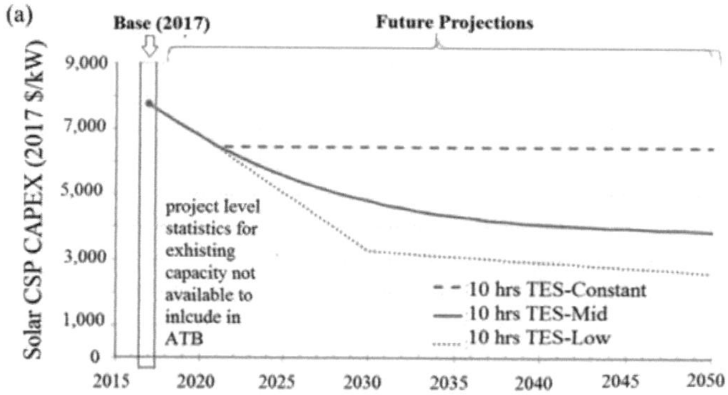

Capital Expenditures (CAPEX): Historical Trends, Current Estimates, and Future Projections for solar CSP.
Source: National Renewable Energy Laboratory Annual Technology Baseline (2019) http://atb.nrel.gov

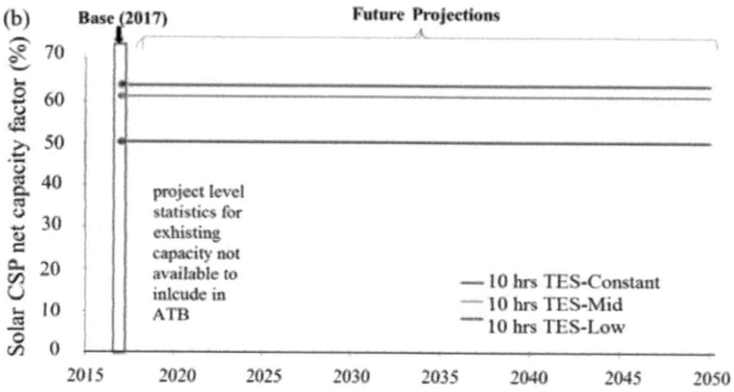

Solar CSP net capacity factor
Source: National Renewable Energy Laboratory Annual Technology Baseline (2019) http://atb.nrel.gov

Figure. 1 - CAPEX and capacity factor of concentrated solar power solar tower with 10 hours of thermal energy storage. Credit NREL.

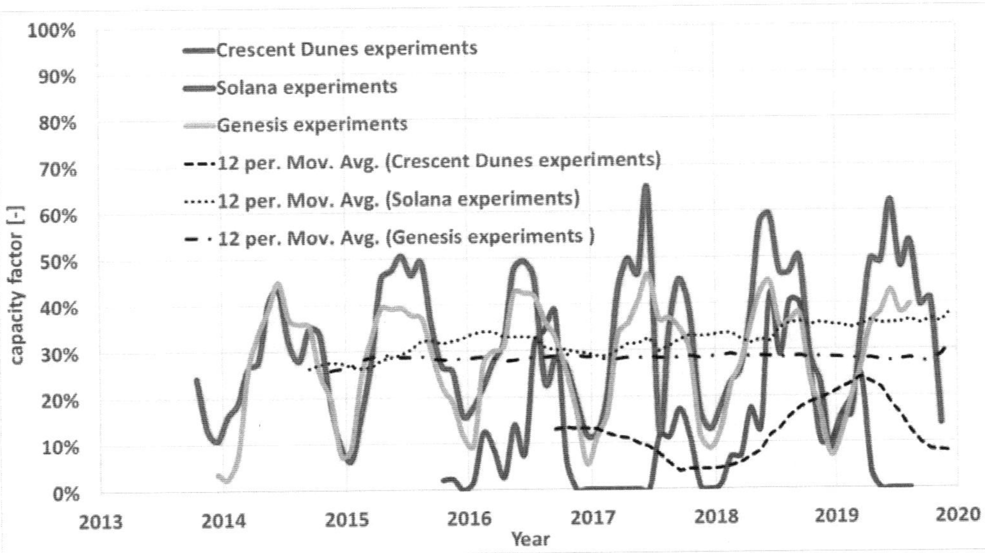

Figure 2 - Comparison of measured capacity factors for Crescent Dunes, Solana, and Genesis.

4. Discussion and conclusions

The best performing large concentrated solar power plant in the world is Solana. It features the more established and reliable parabolic trough technology, plus 6 hours of molten salt thermal energy storage. The 250 MW plant, completed in 2013, had a construction cost of 2 billion $US. This translates into a CAPEX of 8,000 $US/kW, which is less than the 8863 $US/kW of Crescent Dunes. The capacity factor of Solana is also increasing, presently at 36.5%. Working 25 years at this capacity factor would translate into a cost of 0.12 US$/kWh including O&M annual costs. The cost of a plant-like Solana is expected to dramatically reduce as soon as the design could converge towards an industrial product, and some issues especially with the energy storage, that is the most delicate part, could be finally addressed. As shown since the start of operation in 2013, the annual capacity factors of Solana have been continuously improving.

A parabolic trough is a much simpler, reliable, and well-established technology. Thus, good plants can be produced at lower costs. Conversely, the solar tower, especially with the addition of larger thermal energy storage, is a much more difficult and delicate design, still lacking technology readiness, and good plants are hard to be made at reduced costs.

Presently, the best opportunity for concentrated solar power is to further refine the parabolic trough design with molten salt thermal energy storage, up to the definition of an industrial product that if mass-produced could bring to the cost down to the 0.060 US$/kWh. Additionally, modeling should be validated against actual production to provide a reliable forecast, [12],[13].

5. References

[1] atb.nrel.gov/electricity/2019/index.html?t=sc

[2] Boretti A, 2018. Concentrated Solar Power Plants Capacity Factors: A Review. Nonlinear Approaches, in Engineering Applications Energy: Vibrations, and Modern Applications, Liming Dai, Eds., Reza N. Jazar, Springer, New York. dx.doi.org/10.1007/978-3-319-69480-1_2

[3] Boretti A, Castelletto S, and Al-Zubaidy S, 2018. Concentrating solar power tower technology: present status and outlook. Nonlinear Engineering – Modeling and Application (NLENG) 8(1). doi.org/10.1515/nleng-2017-0171

[4] Boretti A, 2018. Cost and Production of Solar Thermal and Solar Photovoltaics Power Plants in the United States. Renewable Energy Focus 26:93-99. doi.org/10.1016/j.ref.2018.07.002

[5] Boretti A, 2019. Realistic expectation of electricity production from current design concentrated solar power solar tower with thermal energy storage. Energy Storage 1(3), e57. dx.doi.org/10.1002/est2.57.

IOP Conf. Series: Earth and Environmental Science **633** (2021) 012003 doi:10.1088/1755-1315/633/1/012003

[6] www.reviewjournal.com/opinion/letters/letter-much-touted-crescent-dunes-solar-plant-goes-bust-1935510/
[7] www.cato.org/blog/crescent-dunes-another-green-flop
[8] nationalinterest.org/blog/buzz/another-federally-backed-solar-energy-project-just-went-belly-116506
[9] www.eia.gov/electricity/data/browser/
[10] sam.nrel.gov
[11] Boretti A, 2020. Cost of dispatchable electricity from concentrated solar power, solar tower plants, with 10 hours' molten salt thermal energy storage. ICACER 2020.
[12] Ezeanya E K, Massiha G H, Simon W E, Raush J R, and Chambers T L, 2018. System advisor model (SAM) simulation modelling of a concentrating solar thermal power plant with comparison to actual performance data. Cogent Engineering 5, 1524051.
[13] Boretti A, Nayfeh J, and Al-Kouz W, 2020. Validation of SAM Modeling of Concentrated Solar Power Plants. Energies 13(8), 1949. doi.org/10.3390/en13081949

IOP Conf. Series: Earth and Environmental Science **633** (2021) 012004 doi:10.1088/1755-1315/633/1/012004

Concentrating solar power multi-effect desalination in Al Khobar, Saudi Arabia

A Boretti[1], S Castelletto[2], W Al-Kouz[1,] and J Nayfeh[1]

1 Mechanical Engineering Department, College of Engineering, Prince Mohammad Bin Fahd University, Al Khobar, Saudi Arabia
2 School of Engineering, Royal Melbourne Institute of Technology (RMIT) University, Bundoora, Victoria, Australia
E-mail: a.a.boretti@gmail.com

Abstract. As about two-thirds of the thermal energy to the power cycle of a concentrated solar power plant is dumped to the condenser, typically air or evaporative cooling, there is the opportunity to achieve energy savings by using a multi-effect distillation system to condensate the steam and at the same time produce desalinized water. Preliminary design considerations and computations for a concentrated solar power plant located in Al Khobar, Saudi Arabia, are here proposed. The solar field has an aperture of 1,230,000 m^2, it uses oil as the receiver fluid, it uses a six hours' thermal energy storage by molten salt, and it features a 140 MW turbine. It is expected production of up to 413 GWh of electricity annually, for a capacity factor of 0.36, by using an air-cooled condenser. Adoption of a glass enclosure and a multi-effect distillation plant replacing the air-cooled condenser may permit the same electricity production plus the production of desalinized water in an environment protected by sand and dust.

1. Introduction

Energy and water supply will be two of the major issues humanity will face during this century. With the projected growth of concentrated solar power (CSP) around the world, the integration of CSP with desalination (CSP+D) has great potential. Often, where there is high solar radiation to support CSP, clean water is even more precious than energy.

Desalination is steam-driven (for example multi-effect distillation, MED) or electricity-driven (for example pressure forced reverse osmosis, RO). Where water conditions give MED an advantage versus RO, and there are the right conditions for CSP, then MED may be conveniently integrated with CSP plants. Regarding CSP, the most promising opportunity lies in the Parabolic Trough (PT) design. Thermal energy storage (TES) by molten salt (MS) introduces the added value of dispatchability but at an increased cost and complexity. CSP with TES is the only renewable energy plant that can deliver power when the resource is not available (wind and solar photovoltaic or CSP without TES need external energy storage by batteries or pumped hydro). CSP with TES features oil as the receiver fluid, then a two tanks TES with MS as the thermal energy storage fluid, and water/steam as the power cycle fluid. The condenser of the Rankine power cycle is air-cooled or evaporative for in-land, arid, desert areas. While there are no examples so far of CSP in coastal locations, it may be convenient to use saltwater once-through condensers. Solar field temperatures are typically 293 K/393 K, and pressure and temperature of the steam to turbine 393 K and 100 bar. This is the design of the 250 MW Solana, in Gila Bend, AZ, the most successful CSP plant presently operated, delivering annual average capacity factors of 36.40% with the opportunity to extend production up to 6 hours after the sunset thanks to the 6 hours of TES. This is also the design of the 250 MW Genesis and Mojave, respectively close to Blythe,

CA and close to Barstow, CA, that do not have TES, and thus only produce electricity during the day. Genesis and Mojave have the latest annual average capacity factors of 28.11% and 24.45%. While the 12 months moving average of the monthly capacity factors is still growing for Solana, this parameter is stable for Genesis and has been recently decreasing for Mojave. Since the start of the operation, the maximum 12 months moving average of the monthly capacity factors was 36.40% in Solana, 29.20% in Genesis, and 28.72% in Mojave.
In areas such as Al Khobar, due to sand, dust, humidity, and proximity to the sea, all the plant equipment has to be enclosed in a glass room. Additionally, the condenser of the power cycle, usually air-cooled or evaporative for the in-land CSP can be made once-trough saltwater condenser. Here we examine the integration of a MED plant with a CSP PT without TES, with the MED being feed by partially expanded steam. Presently, there are in Al Khobar two desalination plants both featuring multi-stage flash distillation (MSF). MSF distills seawater by flashing part of the water into steam in multiple stages of countercurrent heat exchangers. MSF was popular in the past but recently suffers from the competition of RO due to much lower energy consumption. MED has in principle advantages versus MSF, but not an energy consumption advantage compared to RO. The availability of steam from a PT solar field may turn MED into the preferred candidate.

2. MED plants

In desalination, seawater is separated into a freshwater stream with a low concentration of salts and a brine with a high concentration of salts. The most relevant technologies are thermal desalination and membrane desalination. Thermal desalination utilizes heat to change the phase of the seawater from liquid to vapor. Membrane desalination uses pressure to force water through a semi-permeable membrane. While membrane desalination needs the energy to run a pump, that may be produced with any mean, thermal desalination utilizes heat, often in the form of steam, that may be conveniently produced by a CSP plant. MED thermal desalination systems may be competitive with seawater RO under specific water conditions [1]. MED systems are designed with multiple evaporator stages, which evaporate seawater to produce desalinated water.
Normally, in large desalination plants, the number of stages or effects varies between 8 and 16 due to the low-temperature drop per effect. Seawater is heated to approximately 70 °C by a steam supply. Boiling occurs in the sequence of effect vessels, each held at a lower pressure than the last. Because the boiling point of water decreases as pressure decreases, the vapor boiled off in one vessel can be used to heat the next. Only the first vessel (at the highest pressure) requires an external source of heat. The seawater and external steam source both enter the MED system in the first effect on opposite sides of the heat exchanger tubes. Steam is fed to the inside of the tubes, and seawater is sprayed over the outside of the tubes in a typical effect. The seawater, upon contacting the heat exchanger tubes, begins to boil. The vapor produced is collected and is transported to the second effect on the inside of the second effect's heat exchanger tubes. The heating steam from the first effect condenses inside the first effect's heat exchanger tubes and is returned to the power plant as condensate. Concentrated seawater brine in the first effect is collected in trays and transported across the outside of the heat exchanger tubes in the second effect, where again it boils by heating and the pressure change. This process continues throughout the series of effects. After the final effect, the vaporized water that has collected on the inside of the effect heat exchanger tubes is cooled and condensed with seawater in a final heat exchanger.
Most MED designs include a thermal vapor compressor (TVC) to increase system efficiency. A TVC recycles some of the vapor produced in the desalination process to reduce the total amount of steam required to drive the process. A TVC is a steam ejector that entrains low-pressure vapor from a downstream effect with motive steam and discharges the mixture into the first effect. MED plants are generally built-in units of about 100 m^3/day up to 36,400 m^3/day (0.03 to 8 mgd), allowing this design to be utilized in smaller volume applications. Multiple units may be combined in one plant to further increase capacity. The system produces very high-quality product water from the sea or brackish water with a total dissolved solids (TDS) concentration of 25 mg/l or less.

3. CSP plants

A PT is made up of a linear parabolic reflector concentrating the sunlight onto a tubular receiver located along the focal line of the reflector [2]. The tubular receiver is filled with a working fluid that is often oil, molten salt, or even water/steam. The reflector follows the sun. The tracking operates along a single axis. The working fluid is usually heated to 390 to 500 °C, depending on the fluid, oil or molten salt flowing through the receiver. If oil or MS, this fluid is then used in a heat exchanger to produce steam for the power cycle. The shaped mirrors of a PT focus the sunlight on a tube running along the focus line with typically an 80x concentration. The sunlight is absorbed by the tube in a glass vacuum and delivered to the receiver fluid. While PT may be less efficient than a solar tower (ST), where a huge number of two-axis tracking heliostats focus the solar energy on a small receiver atop of a tall tower, they are much simpler and less expensive to build and operate. Reference PT specifications change if the fluid is water/steam, oil, or MS as follows, ref. [2] to [8].

For oil as the receiver fluid, the receiver temperature is 390 ºC. The peak flux on the receiver is 25 kW/m^2. The hot storage temperature if any is 390 ºC, the cold storage temperature is 290 ºC, and the condenser temperature is 40 ° C for heat rejection. This is the configuration used for example in the CSP PT plant of Solana, Genesis, and Mojave.

For MS as the receiver fluid, the receiver temperature may go up to 500 ºC, the peak flux on the receiver is 25 kW/m^2, the hot storage temperature is 500 ºC, and the cold storage temperature is 300 ºC. The condenser temperature is unchanged. There is no large CSP PT plant presently featuring this opportunity.

In the case of water/steam as the receiver fluid, the receiver temperature may also go up to 500 ºC, the peak flux on the receiver is 25 kW/m^2. The hot and cold storage tanks are not available in this case, all the other parameters being unchanged. Similarly, there is no large CSP PT plant presently featuring this opportunity.

Supercritical steam power cycles and supercritical carbon dioxide power cycles are also being considered to improve the conversion efficiency thermal-electric. Higher temperature receiver fluids, and materials and manufacturing improvements to reduce costs are also the major area of concern for CSP PT.

Enclosed troughs are used to protect the plant equipment from high winds and blowing sand and dust [9]. Glass Point solar steam generators [9] house thin curved mirrors inside a greenhouse. The mirrors track the sun throughout the day, focusing the heat on pipes containing water. The concentrated sunlight boils the water to generate steam. The greenhouse has an automated washing machine to maintain optimal performance even in harsh conditions. The Glass Point PT system is only used to produce steam then used in the oil fields.

In Saudi Arabia, in coastal areas such as Al Khobar, sand, and dust, plus humidity and salinity for the proximity to the sea, suggest enclosing all the plants in a glass room. In the case of a plant producing electricity, it is also important to adopt a once-through seawater condenser. In this case, the most reliable technology is oil as the receiver fluid, then one heat exchanger oil to steam/vapor. If the production of electricity is the main focus of the plant, the best opportunity is then to adopt thermal energy storage by molten salt. If the main focus of the plant is to produce desalinized water, then the best option is to use water as the receiver fluid and then MED. If the focus of the plant is to produce both electricity and desalinized water, then the best option is again to use water as the receiver fluid, with then part of the steam to the turbine diverted to the MED, or also adopt the classic scheme with oil as the receiver fluid, a heat exchanger to produce steam, and then use part of the steam to the turbine diverted to the MED.

4. Electricity plus desalination

A key issue with deploying renewable energy conversion systems is their intermittent nature. Solar photovoltaics (PV) generate power only when the sun shines. CSP, if fitted with TES, may theoretically produce electricity 24/7, even if in practice CSP plants with TES have about the same capacity factors of CSP plants without TES, much higher costs, and reduced reliability, as the technology is everything but mature. The 6 hours of thermal energy storage in Solana permit power generation after sunset during the summer months, up to the early hours of the following morning, as it is otherwise impossible without TES. The benefits of energy storage are invaluable in improving grid stability, power quality, and the continuity of supply. However, TES is difficult to be implemented.

Combining CSP with desalination has an energy storage advantage. Many solar fields for CSP are designed to match an average insolation flux. In the summer and during peak daylight hours, there is excess heat produced by the solar field. In the absence of TES, some of the solar collectors or mirrors must be defocused (energy dumping). This excess heat may be used in the desalination plant. In principle, as the power demand seasonally changes, a combination of solar power generation and desalination may allow some additional degree of discretionary allocation of the heat input to the production of power or water desalination. CSP plants usually have a steam power cycle where expansion in the turbine is limited by the conditions to the air-cooled or evaporative condenser. In hot locations, a once-trough seawater condenser provides better efficiencies of the power cycle than an air-cooled or evaporative condenser. In the specific area, an air-cooled or evaporative condenser is difficult to be used also for other reasons, from the salt to the sand and dust to the humidity. Using a MED as condenser may also provide advantages vs. an air-cooled or evaporative condenser.

A CSP plant without TES plus a steam/water power cycle coupled to a switchable MED desalination plant/once trough seawater condenser may give advantages vs. a traditional CSP plant without TES with an air-cooled/evaporative condenser, producing desalination in addition to more electricity.

A CSP plant without TES coupled to MED is shown in ref. [10]. The Thermoflex [11] model describes a plant in the coastal Surt region of Libya producing up to 15,000 m^3per day of potable water while also delivering electricity up to 15 MW of power. The receiver fluid is water/steam. The plant, which also features a natural gas combustor to support the solar field in the production of steam, has no air-cooled or evaporative or once-through saltwater condenser, but only the MED plant. The hybrid design fossil fuel – solar can produce power and water 24/7 without thermal energy or other external storage such as batteries or pumped hydro. The MED plant replaces the condenser to also produce potable water from seawater. The solar field uses Linear Fresnel Collectors (LFC) with direct steam generation (DSG) rather than the parabolic trough. Turbine inlet steam conditions are 55 bar and 400 °C. The MED design has a nominal performance ratio of 10.4 by using 12 effects over a temperature range of 30 °C. Thermoflow/thermoflex is the preferred software for this kind of modeling, permitting the accurate representation of the turbine sections, the condenser, and the many heat exchangers involved in the plant design.

Many other options are certainly possible. Solar field plus a natural gas burner in parallel, or solar field only, or solar field plus thermal energy storage to produce steam, and then one traditional condenser once trough saltwater, one condenser once trough saltwater and the MED, or the MED only, to condense the steam. In between these other alternatives, a purely theoretical CSP plant with TES coupled to MED is shown in ref. [12]. The model is based on Transient System Simulation Tool (trnsys) [13], which is based on what concerns the CSP on SAM [14]. The receiver fluid is oil. The heat storage fluid is MS. Then, there is a heat exchanger for the MS and a heat exchanger for the water/steam. The plant does not feature a boiler to support the solar field in the production of steam. It has no air-cooled or evaporative or once-through saltwater condenser, but only the MED plant to serve as condenser.

5. Preliminary simulations

A SAM [14] model was developed for half of the Solana plant in Gila Bend, with experimental and computational capacity factors compared in the validation exercise. Solana has 6 hours of molten salt thermal energy storage and air-cooled condenser. As the real plant still suffers from maturity issues with 12 months moving average of the capacity factor always increasing (latest 36.4%), operation of many components is overrated in SAM that is proposing a 41% annual average capacity factor over the typical year. Fig.1 presents a comparison of the computed and measured capacity factor for Solana, Gila Bend, AZ. Fig.2 then presents the computed capacity factors for a plant the same as one half of Solana, except a 10% larger specific solar field, located in Al Khobar, Saudi Arabia. Fig. 3 finally presents the power of collecting energy in the solar field, the power to and from the storage, and finally the power to the steam power cycle. As the rated efficiency of the power cycle is about 36%, about 2/3 of the thermal energy from the power cycle gets dumped to the condenser.

Unfortunately, SAM has a very crude representation of the turbine and condenser, and it does not permit to model once-trough saltwater condensers or MED plants. The SAM condenser pressure is dependent on ambient air temperature, cooling technology, and cooling load through simplified relations. The only two cooling technologies available are wet cooling and dry cooling. Both wet and dry cooling use ambient air as the cold thermal reservoir. Wet cooling uses the wet-bulb air temperature. Dry cooling uses the dry-bulb temperature of the air.

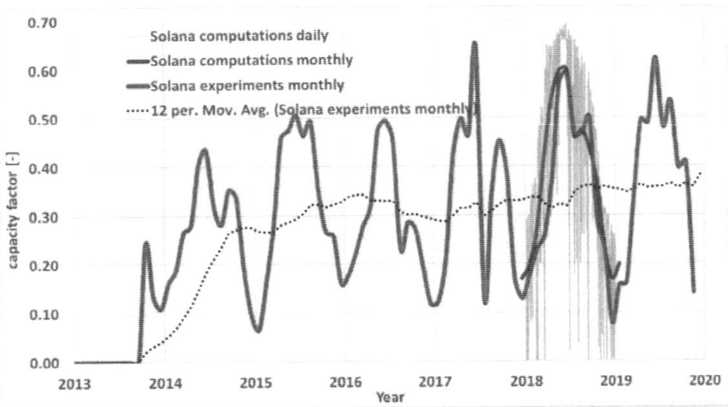

Figure 1. The experimental and computational capacity factor for Solana, Gila Bend, AZ. Computations were performed using SAM.

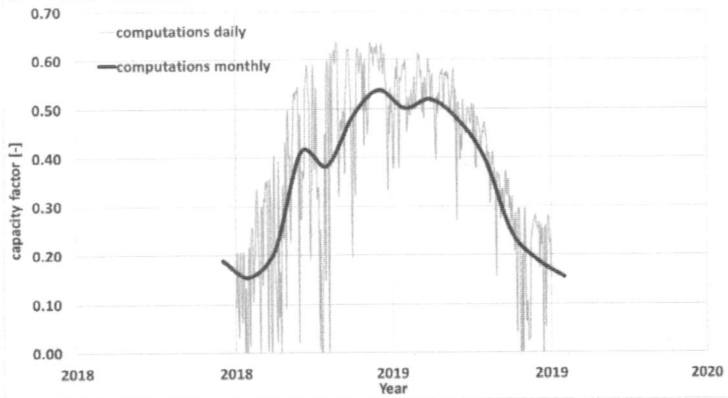

Figure 2. The computational capacity factor for a plant like ½ of Solana, with the specific solar field, increased 10%, located in Al Khobar, Saudi Arabia. Computations were performed using SAM.

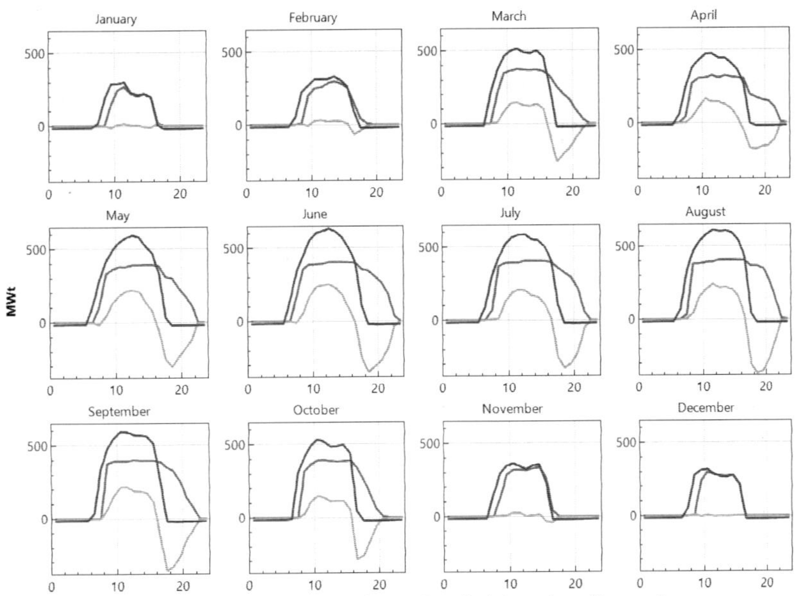

Figure 3. Computational thermal energy from the solar field and to/from the storage, and thermal power to the power cycle, over the 24 hours of a typical day of every month for the Al Khobar plant. Computations were performed using SAM. About 2/3 of the energy to the power cycle is then dumped to the condenser. This energy may be used for MED.

IOP Conf. Series: Earth and Environmental Science **633** (2021) 012004 doi:10.1088/1755-1315/633/1/012004

The dry-bulb temperature can be significantly higher than the wet-bulb temperature. With a once trough seawater condenser, the temperature of the cold thermal reservoir is otherwise the water temperature, that especially daytime during the summer is much lower than the dry-bulb air temperature. With MED distillation, the temperature of the cold thermal reservoir is larger than the water temperature but difficult to say if there are gains or losses versus the air-cooled condenser without performing appropriate simulations with better software tools such as thermoflow.

6. Conclusions

In coastal locations, where sand, dust, humidity, and salt also play a major role in concentrated solar power plants operation, there is the convenience to adopt a glass enclosure for the parabolic trough plant components, as well as to use a once trough saltwater condenser in place of traditional air-cooled or evaporative cooling condensers. As two-thirds of the thermal energy to the power cycle gets dumped at the condenser, an interesting opportunity is to use a multi-effect distillation plant as the condenser of the steam power cycle. This way, the plant produces both electricity and desalinized water. Following this preliminary exercise, better simulations are needed to determine the feasibility of such a plant in terms of costs and performance. For a 24/7 operation, a natural gas or oil burner can be conveniently added to the basic concentrated solar plant design.

7. References

[1] Zachary J and Layman C M, 2010. Adding Desalination to Solar Hybrid and Fossil Plants, powermag.com, www.powermag.com/adding-desalination-to-solar-hybrid-and-fossil-plants/.
[2] Boretti A, 2018. Concentrated Solar Power Plants Capacity Factors: A Review, Nonlinear Approaches in Engineering Applications Energy: Vibrations, and Modern Applications, Liming Dai, Eds., Reza N. Jazar, Springer, New York.
[3] Margolis R, Coggeshall C, and Zuboy J, 2012. SunShot vision study. US Dept. of Energy.
[4] Radosevich L G,1988. Final Report on the Power Production Phase of the 10 MWe Solar Thermal Central Receiver Pilot Plant, SAND87-8022. Sandia National Laboratories, Albuquerque, NM.
[5] Pacheco J E, Wolf T, and Muley N, 2013. Incorporating Supercritical Steam Turbines Into Advanced Molten-Salt Power Tower Plants: Feasibility and Performance. Sandia National Laboratories, Albuquerque, NM, Report No. SAN2013-1960.
[6] Kearney D and Herrmann U, 2006. Engineering evaluation of a molten salt HTF in a parabolic trough solar field. National Renewal Energy Laboratory.
[7] Feldhoff J F, Schmitz K, Eck M, Schnatbaum-Laumann L, Laing D, Ortiz-Vives F, and Schulte-Fischedick J, 2012. Comparative system analysis of direct steam generation and synthetic oil parabolic trough power plants with integrated thermal storage. Solar Energy, 86(1):520-530.
[8] Bendt P, Rabl A, Gaul H W, and Reed K A, 1979. Optical analysis and optimization of line focus solar collectors (No. NREL/TR-34-92). National Renewable Energy Laboratory (NREL), Golden, CO.
[9] www.glasspoint.com/technology/
[10] www.thermoflow.com/images/Solar%20Thermal%20Pamphlet%202012.pdf
[11] www.thermoflow.com
[12] Casimiro S, Cardoso J, Alarcón-Padilla D C, Turchi C, Ioakimidis C, and Mendes J F, 2014. Modeling multi effect distillation powered by CSP in TRNSYS. Energy Procedia 49:2241-2250.
[13] www.trnsys.com/
[14] sam.nrel.gov

IOP Conf. Series: Earth and Environmental Science **633** (2021) 012005 doi:10.1088/1755-1315/633/1/012005

The energy future of Saudi Arabia is not batteries and photovoltaic solar panels

A Boretti[1], S Castelletto[2], W Al-Kouz[1,] and J Nayfeh[1]

1 Mechanical Engineering Department, College of Engineering, Prince Mohammad Bin Fahd University, Al Khobar, Saudi Arabia
2 School of Engineering, Royal Melbourne Institute of Technology (RMIT) University, Bundoora, Victoria, Australia
E-mail: a.a.boretti@gmail.com

Abstract. Recently, North European academics claimed that "Saudi Arabia can accomplish a 100% renewable energy by 2040 that is overwhelmed by PV single-axis following and battery storage". They moreover say "Battery capacity contributed up to 30% of the full power request in 2040 and the commitment increments to 48% by 2050". Based on geology, climate conditions, and resources of Saudi Arabia, it is clarified as batteries and photovoltaic solar panels are not the best choice for the country's energy sector. To cover all the energy supply of Saudi Arabia by solar photovoltaic, plus battery capacity to compensate for the solar energy fluctuations, and unpredictability, is not convenient, for both the economy and the environment. Superior results for the environment and the economy may be accomplished by valorising the fossil fuel assets, through the development of high-efficiency plants such as the combined cycle gas turbine plants of Qurayyah, the advancement of innovations for the generation of clean power, counting oxyfuel combustion and carbon capture and storage, and also developing nuclear energy. The development of nuclear power plants may be more advantageous to the economy and the environment than photovoltaic and batteries. Concerning solar energy, parabolic trough solar thermal power plants also along the coast have advantages versus solar photovoltaic, as thermal energy storage could provide dispatchability. Additionally, a centralized power plant works better than many rooftops photovoltaic that will be likely covered by sand, rusted, or cracked after a few years.

1. Introduction

A North European scholastics article [1] titled "The role that battery and water storage play in Saudi Arabia's transition to an integrated 100% renewable energy power system" has claimed that "... Saudi Arabia can accomplish a 100% renewable energy by 2040 by utilizing PV single-axis and battery storage. Single-axis PV contributed 210 GW out of the whole 403 GW by 2040. The commitment expanded to 369 GW out of an add up to 520 GW by 2050. Battery storage contributed up to 30% of the full power request in 2040 and the contribution increases to 48% by 2050." Apart from using past tense for the coming years, batteries don't produce power. They only store the power delivered by something else to make use of when needed. Moreover, battery storage capacity is still in its infancy, and not demonstrated off the shelf innovation, and it has critical financial, natural, and maintainability that don't allow their spreading worldwide. Concerning the source of the power that the batteries may store for future discharge, it is difficult to accept that this power can be created by solar photovoltaic (PV) only. They are much less productive and much costlier than what it is depicted. Furthermore, they endure particular issues in Saudi Arabia such as sand that significantly increments the costs and decreases efficiency over time. The lifetime of solar panels in Saudi Arabia is definitively less than the 20 years that are usually assumed in the power purchase agreement of other countries.
Contrary to what it may be upheld by articles such as ref. [1] claiming that "Saudi Arabia can accomplish a 100% renewable energy by 2040 overwhelmed by PV single-axis and battery storage", based on real-world data from the countries that publish cost and electricity production results, we clarify why this suggestion is impracticable and not helpful.

PV solar panels still have critical costs and work with typical capacity factors of around 0.27-0.28 in better situations such as Australia. This is the case of more expensive (per kW of power) energy facilities. The capacity factor can be as low as 0.10 in the event of rooftops [2], [3], [4], [5]. In Saudi Arabia, sand and dust radically diminishes efficiency as well as the lifetime and require costly cleaning frameworks.

The capacity factors of individual facilities moreover are affected by variability, with a standard deviation of high-frequency distributions largely above the mean over one year [2], [3], [4]. Indeed, at the grid level, variability is large, as solar energy is never accessible at night time and it also changes during the day and with the season.

Battery storage is in its infancy. The world's biggest battery, installed in Australia, only has 100 MW of nominal power and 140 MWh of nominal energy storage [2]. It is never charged or discharged over 30% of nominal power. Similarly, charging and discharging are not from 0 to 100% state-of-charge but only a fraction of this.

Based on their life cycle appraisals, the use of well-established combined cycle gas turbines plants is better, from financial, natural, ethical, sustainability, point of view.

In case Saudi Arabia is focusing on a better environment and low carbon energy economy, it would be way better to contribute in nuclear energy for baseload, whereas cultivating more productive utilize of fossil powers through high-efficiency plants such as combined cycle gas turbine plants. Cleaner employments of fossil fuels [6] is additionally an opportunity to investigate.

Concerning solar, parabolic trough solar power created along the coast is more promising than solar PV.

Concerning the opportunity to utilize pumped hydro energy storage, Saudi Arabia has no lakes nor rivers. [1] claims, "The role that battery and water storage play in Saudi Arabia's transition to an integrated 100% renewable energy power system". It must be recalled that Saudi Arabia has no rivers and exceptionally little water. Whereas conventional hydropower in other nations may certainly be modified to pumped hydro by including pumping capabilities [2], Saudi Arabia has no existing hydropower facilities. Coastline pumped hydro is still in its infancy, a single little pilot plant was built, run for few years, and decommissioned in Japan. Not working technologies should not be exaggerated to diminish well-proven technologies.

Here we show the operation of real-world PV and wind within Australia where the high-frequency information about electricity production is made accessible to the public. The Australian national electricity market (NEM) is an open framework where operators and traders purchase and offer electricity with straightforwardness. The Australian NEM information is also discussed in [7].

The information about energy generation from ref. [7] is accessible for every 5 minutes – day by day information – or 3 hours – month to month information, in terms of power or capacity factor. The capacity factor is the proportion of the real producing power to the enlisted capacity. The total enlisted capacity of sun-powered facilities (all PV) is 3,427 MW. The information of ref. [8] is accessible every 30 minutes for the everyday charts and every 3 hours for the month to month charts. The total registered capacity of the sun-powered rooftop is almost 8,000 MW. The information of ref. [9] is accessible every 5 minutes – daily data – or 3 hours – month to month information, as power or capacity factor. The full enlisted capacity of battery energy storage is 191 MW.

2. Results

High-frequency information of solar and wind power generation, climate conditions, and power plant components and system yield is fundamental to plan the energy capacity required to create a stable grid supplied by irregular and unpredictable wind and solar-based energy. To precisely validate and make solid renewable energy software tools, they must be based on high-frequency data (each minute or less) for climate and plant operation.

Fig 1 presents sample results for the Australian National Electricity Market grid, managed by the ANEMO. The capacity factor of wind and solar energy facilities is shown during one sample month and two sample days. The variability is dramatic also at the complete grid level. This requires a huge installed capacity to deliver the energy needed by the grid as well as huge storage to compensate between peaks and valleys.

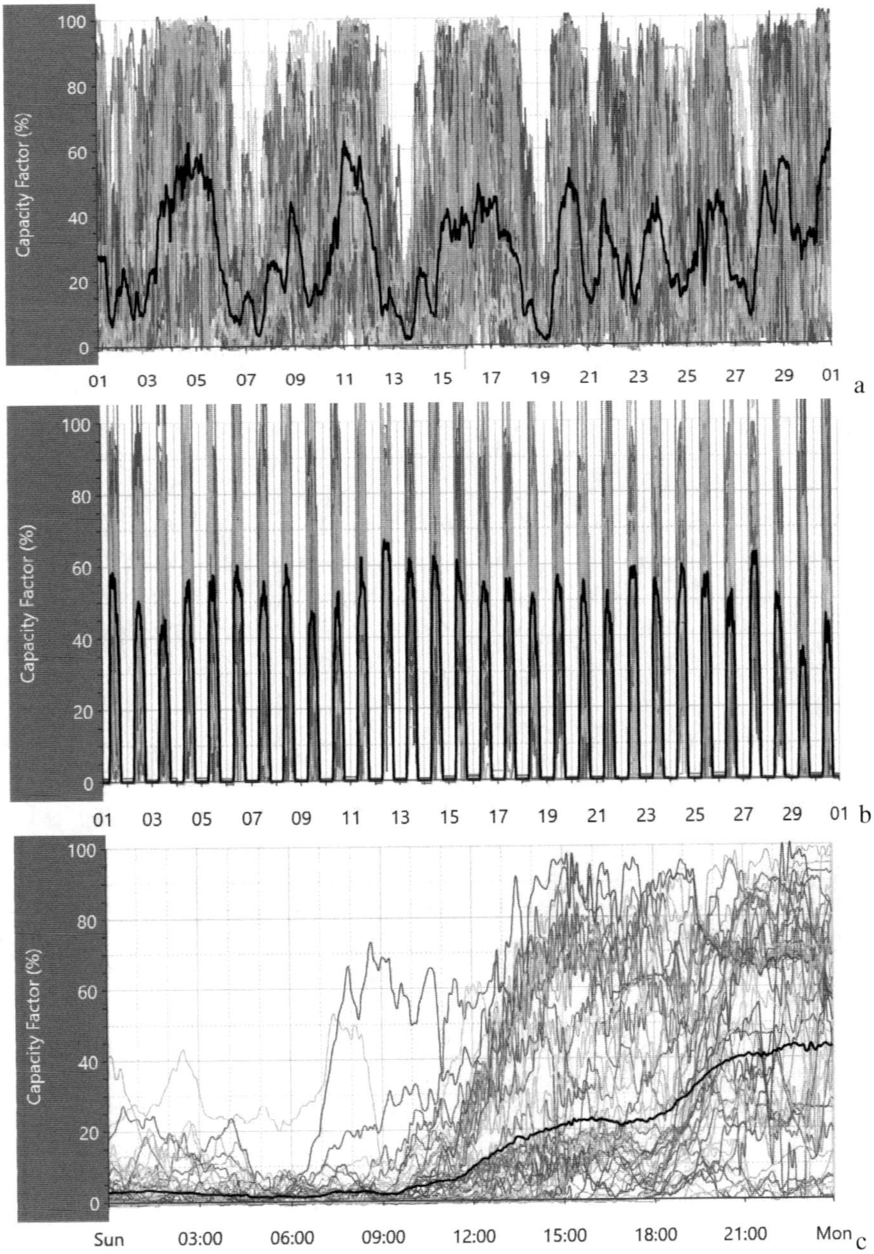
Capacity Factor (%)
100
80
60
40
20
0
01 03 05 07 09 11 13 15 17 19 21 23 25 27 29 01
a
Capacity Factor (%)
100
80
60
40
20
0
01 03 05 07 09 11 13 15 17 19 21 23 25 27 29 01
b
Capacity Factor (%)
100
80
60
40
20
0
Sun 03:00 06:00 09:00 12:00 15:00 18:00 21:00 Mon
c

IOP Conf. Series: Earth and Environmental Science **633** (2021) 012005 doi:10.1088/1755-1315/633/1/012005

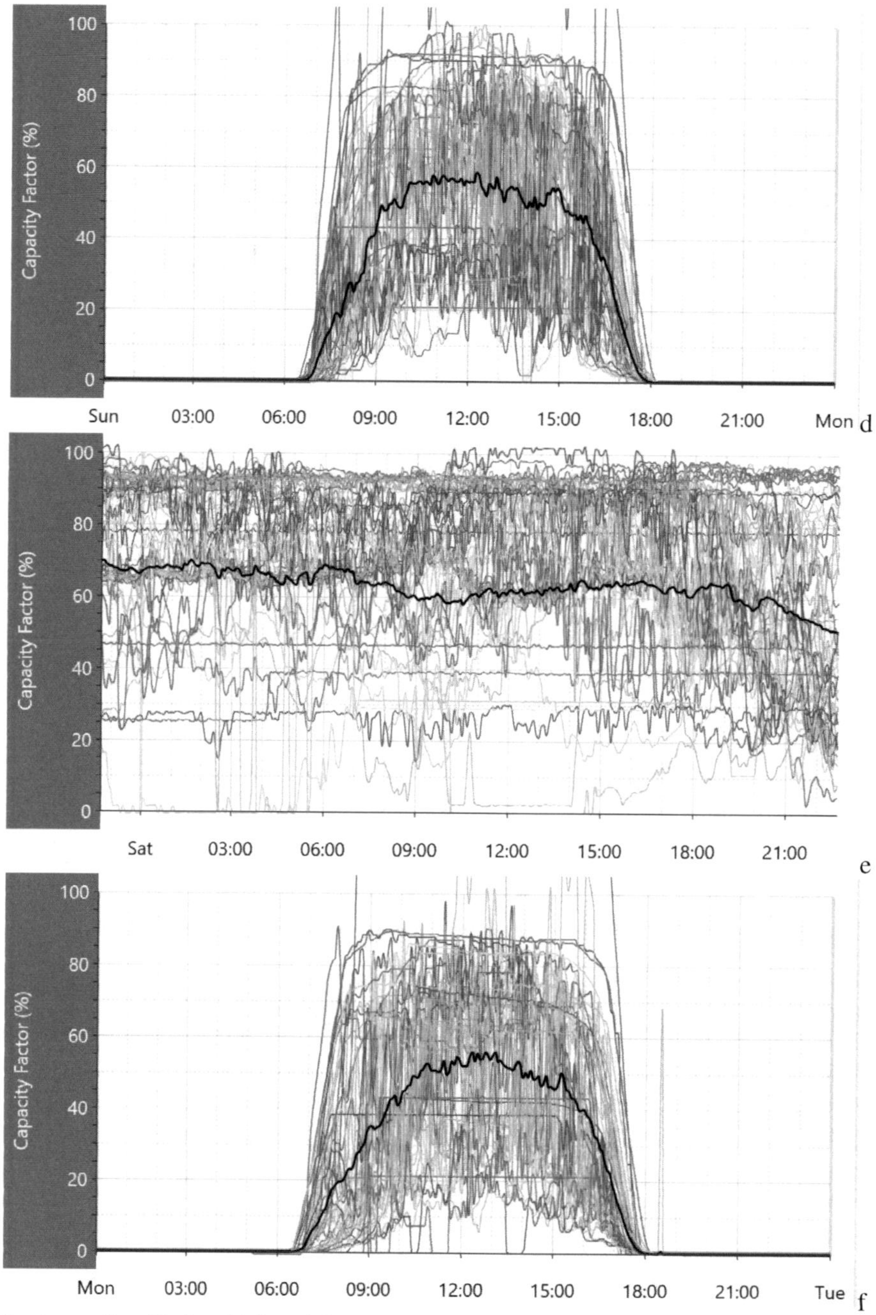

Figure 1. Australian National Electricity Market grid, managed by the ANEMO. (a) The capacity factor of wind energy facilities during April 2020, Individual facilities and grid average. (b) The capacity factor of solar energy facilities during April 2020, Individual facilities and grid average. (c) The capacity factor of wind energy facilities during 19 April 2020, Individual facilities and grid average. (d) The capacity factor of solar energy facilities during 19 April 2020, Individual facilities and grid average. As some facilities are registered before being fully completed, their nameplate capacity may change over the year, and the registered capacity is updated with a delay, some capacity factors may exceed 100. Sunday 19 April before 6 AM very little renewable energy was supplied by wind and solar. (e) The capacity factor of wind energy facilities during 20 April 2020, Individual

facilities and grid average. (f) The capacity factor of solar energy facilities during 20 April 2020, Individual facilities and grid average. The night after, wind energy facilities were operating at a 70% capacity factor rather than the less than 3% capacity factor of the day before. Images reproduced modified from [10].

3. Conclusions

The only two countries on earth providing transparent data of renewable energy are the United States and Australia. The United States Energy Information Administration (EIA) publishes the actual electricity production of every plant in the country, renewable or not renewable, every month, since 2001, or the registration of the plant if later. The Australian Energy Market Operator (AEMO) publish the actual electricity production data of every plant in every state of the country excluded Northern Territory and Western Australia.

The low-frequency data show a capacity factor of about 30-33% of large PV or CSP solar, as well as wind energy facilities, and significant seasonal variability. The high-frequency data show a much more dramatic variability not only at the single plant level but also at the full grid level. During nights of low wind, no matter larger may be the installed capacity, the electricity production is proven to be negligible. This call for a proper reassessment of the capacity of solar and wind needed, based on the actual power of generating power across the year, as well as the power and the energy of the massive storage needed to make a power grid overwhelming wind and solar balanced.

In terms of average performances over a year, the 250 MW CSP parabolic trough of Solana, with limited dispatchability thanks to 6 hours of thermal energy storage, has a capacity factor of 36% at the cost was 2 billion $. Without thermal energy storage, and no dispatchability 250 MW CSP parabolic trough of Genesis has a capacity factor of 28% at the cost of 1.25 billion $. The first plant is expected to work for 20-25 years, the second up to 30 years (examples of similar SEGS plants built in the 1980s). The Crescent Dunes solar tower with thermal energy storage of cost 1 billion $ for 110 MW has been already shut down after only 4 years for lack of production (it has produced in the 4 years less than what was expected in every single year).

Costs are better with PV, but with life not expected to last more than 20 years (and very likely they last much less) and zero dispatchability. Solar Star had for 597 MW a cost of 2.5 billion $ for a capacity factor of 32.5%. The less performing desert Sunlight had for 563 MW a reported cost, likely underrated of 1.46 billion $ for a capacity factor of 26% and the same technology Topaz for 585.9 MW had a cost of 2.50 billion $.

Regarding wind, of life only 20 years, the Roscoe Wind energy facility of power 781.5 MW had a cost above 1 billion $ for a capacity factor of 32%. The units I to IX of Alta Wind, totaling 1,320 MW, had a cost of 2.875 billion $ for a capacity factor of 25%. The 496.6 MW Bison Wind Energy Center had a cost of 0.8 billion $ for a 33% capacity factor. There is no reason to expect better performances for wind and solar plants in Saudi Arabia than in the United States or Australia, due to the extreme temperatures plus sand and dust that are particularly challenging.

The Qurayyah IPP gas and oil-fired combined cycle power station in Qurayyah, Saudi Arabia had a cost of 2.85 billion $ for 3,927 MW. The plant was properly designed to cope with sand and dust and extreme temperatures. Experience from the United States tells us that the natural gas combined-cycle power plants operated in the 50% to 80% capacity factors range in 2015. Considering the fuel conversion efficiency is 52%, and the electricity production is completely controllable rather than fully unreliable as per wind or solar, the cost of oil and natural gas, and the availability of oil and natural gas, there is no doubt that the present and near future of Saudi Arabia must be gas and oil-fired combined cycle power stations. Regarding the life span of these plants, the Rabigh-2, a two 413-MW, combined-cycle blocks using oil, of 826 MW of total power was completed in 1988 and 1993, and it is well working in 2020 after 32 and 27 years respectively, without being cracked or rusted.

The real cost of dispatchable electricity by solar PV plus batteries based on real-world experiences is summarized below, by using the cases of Solar Star and Topaz for what concerns the solar energy by PV without storage, and the latest Hornsdale expansion for what concerns the battery storage. Solar PV in Saudi Arabia is expected to produce less energy and have a shorter lifetime than in California. Similarly, batteries in Saudi Arabia are expected to have reduced efficiency and lifetime when compared to batteries in South Australia. The cost of the battery storage of the Hornsdale expansion is 0.77 million $US per MWh (71 million AU$ equivalents to 49 million US$ for 64 MWh).

Solar star 1 and 2 have a nominal power of 579 MW and a capacity factor of 0.325. This means the actual power is 188 MW. The cost was 2.5 b$. Neglecting O&M cost and assuming 30 years' lifespan, the cost of electricity is about 0.05 $/kWh (≈2.5e9/188e3/30/365/24). The minimum battery requirement is 188x12 MWh actual or 188x12x2 MWh nominal. If we take the life span 30 years and we do not consider O&M costs, the additional storage cost is about 3.75 b$ (≈0.77x188x12x2/1000). This is optimistic. If we take the life span 15 years and similarly we do not consider O&M costs, the additional storage cost is about 7.0 b$. The total cost inclusive of storage is thus about 0.19 $/kWh.

About the same numbers are given by Topaz. Topaz has a nominal power of 589.5 MW and a capacity factor of 0.25. Thus, the actual power is 147 MW. The cost was 2.5 b$.

Neglecting O&M cost and assuming 30 years' lifespan, the electricity cost is about 0.065 $/kWh (≈2.5e9/147e3/30/365/24). The minimum battery storage needed is 147x12 MWh actual, 147x12x2 MWh nominal.

By assuming a life span of 30 years and no O&M costs, the additional cost is about 2.7 b$ (≈0.77x147x12x2/1000). By assuming a more realistic life span of 15 years and no O&M costs, the additional cost is about 5.4 b$. The total cost inclusive of storage is thus about 0.20 $/kWh.

As solar energy has a strong correlation, this is roughly the cost that high-efficiency solar PV plant technologies would have to power for example NEOM City.

The current state of renewable energy is well represented by ref. [11], and it is very far from what is portrayed. Opposite to the present or the near future, in a future not that close there is certainly the need for a mix of solar, wind, and novel nuclear power plants, with all of them requiring significant investment in research and development, to reduce the share of oil and natural gas. Especially wind and solar must be made better, producing more and costing much less, for both the economy and the environment. More specifically, the future of Saudi Arabia by 2040 can't be PV solar panels and batteries. Their economic and environmental costs will be unaffordable, as with present technologies they will damage even more the environment versus the present situation wasting huge amounts of money.

4. References

[1] Caldera U, and Breyer C, 2018. The role that battery and water storage play in Saudi Arabia's transition to an integrated 100% renewable energy power system. Journal of Energy Storage, 17:299-310.

[2] Boretti A, 2019. Energy storage needs for an Australian National Electricity Market grid without combustion fuels, Energy Storage. Energy Storage. DOI:10.1002/est2.92

[3] Boretti A, 2019. High-frequency standard deviation of the capacity factor of renewable energy facilities - part 1: solar photovoltaic, Energy Storage. DOI:10.1002/est2.101

[4] Boretti A, 2019. High-frequency standard deviation of the capacity factor of renewable energy facilities - part 2: wind, Energy Storage. Energy Storage 1(6), e100. DOI:10.1002/est2.100

[5] Boretti A, 2018. Cost and Production of Solar Thermal and Solar Photovoltaics Power Plants in the United States. *Renewable Energy Focus*. **26**:93-99. doi.org/10.1016/j.ref.2018.07.002

[6] Boretti, A. 2020. Perspectives of production of hydrogen for export from wind and solar energy, natural gas, and coal in Australia, International Journal of Hydrogen Energy, International Journal of Hydrogen Energy 45(7):3899-3904.

[7] anero.id/energy/solar-energy/

[8] anero.id/energy/rooftop-solar-energy

[9] anero.id/energy/battery-storage-energy

[10] anero.id/energy

[11] Moore, M., 2020. Michael Moore Presents Planet of the Humans, Full Documentary, Directed by Jeff Gibbs. www.youtube.com/watch?v=Zk11vI-7czE

IOP Conf. Series: Earth and Environmental Science **633** (2021) 012006 doi:10.1088/1755-1315/633/1/012006

Analysis of electricity production through photosynthesis in tropical climate plants

Steven Cedeño[1], Kimberly Mora[1], Lata-García Juan[1]

1 Universidad Politécnica Salesiana, Sede Guayaquil, Ecuador
Facultad de Ingeniería Eléctrica, Sede Guayaquil, Ecuador
GIPI, Grupo de investigación de Procesos Industriales
E-mail: scedenop@est.ups.edu.ec, kmoras1@est.ups.edu.ec, jlatag@ups.edu.ec

Abstract. A modern and competent world demands technological innovation in all areas, in the field of electricity, it has gained impressive strength since the development and implementation of clean energy sources, with advances in energy generation that we could hardly have imagined in the last century, but the need for consumption is increasing and the integration of systems capable of producing electricity in a sustainable and economic manner has never been more imperative. The use of a prototype biological cell composed of an external circuit and configured in such a way that it is possible to obtain levels of electrical energy, presents a future option for the self-sustainable development of electricity, everything is carried out together with the photosynthetic process of the plants, in the studied configuration, the oxidation of organic molecules release protons and electrons that travel through the anode and cathode properly integrated, resulting in a flow of energy easily measurable.

1. Introduction

The world's electrical panorama is directed towards energy efficiency, self-sustainability and technological innovation. Every year, international organizations bet on clean energy alternatives, and this thanks to decisive factors, such as low implementation costs, high rates of electrical productivity, and reduction of CO2 emissions.

Ecuador currently maintains a very significant participation in the framework of renewable energies. According to official data from ARCONEL, in 2018, the nominal power at the national level was 8,676.89 MW; of which 5,271.74 MW (60.75%) corresponded to plants with renewable energy sources and 3,405.14 MW (39.24%) to plants with non-renewable energy sources [1], [2]. It is a favorable point for the Ecuadorian electrical sector, a large part of these advances go hand in hand with the restructuring of strategic sectors by the Ecuadorian government, advances such as these give room for research and development of non- conventional technologies for electrical production. This research work is based on previous studies related to biomass, using living plants in microbial fuel cells (P- MFC) study introduced and applied in 2012. The system is made up of cells that act together with the natural process of photosynthesis carried out under the incidence of sunlight, more than half of the organic matter is released in the roots initiating an action of bacterial oxidation typical of this cell, resulting in a release of electrons and protons, conducted through the anode, the electrons flow due to the potential difference, from the anode through an electrical circuit with cathode resistance, as a result there is an electricity.

An attractive approach to storing solar energy in chemical bonds is photosynthesis, however it has been a challenge in recent decades to build semi-artificial photosynthetic systems that are efficient at converting solar energy and storing it in the form of chemical bonds of solar fuels. Reference [3]

reviews in detail the molecular level of natural photosynthesis, further demonstrating that artificial photosynthesis is composed of light gathering and charge separation units along with catalytic units of oxidation and reduction of water as well as CO2.
In figure 1, we can see a descriptive diagram of the operation of a microbial fuel cell in 2008, a new renewable-type technology was introduced especially biomass known as the plant microbial fuel cell (P- MFC). Electricity can be generated from living plants and therefore bioelectricity and biomass production can be combined on the same surface, in reference [4] using granular activated carbon for the anode, gold wires as current collector while Teflon-coated copper wires are the connecting anode and cathode.

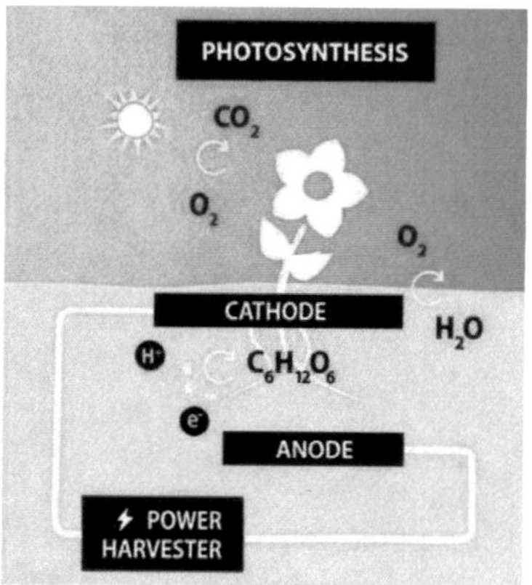

Figure 1. Descriptive diagram of microbial fuel cell.

Reviewing, studying the state of the art [5], [6], [7], [8] leads us to the creation of a functional prototype at the beginning of the fuel cell with an internal circuit in addition to the anode and cathode capable of conducting the energy, we adapted the system to a controlled space where we determined the variables such as temperature, soil composition, levels of material conditions, physical effects of metals, in search of a feasible alternative it was decided to carry out a performance analysis [9].
Where, a key factor is the selection of the plant species, 5 different species are used that can be commonly found in the coastal region, and are used for different purposes such as decoration, feeding or medicinal. Naturally, all plants carry out the process of photosynthesis in such a way that there are species that present a greater efficiency in oxidizing organic waste and producing electrons, this being determined by native applications of similar systems where the variety is summarized in two species [10].
It can be pointed out that studies mark a precedent and diverse approaches throughout its integration, it was demonstrated that the voltage levels are low to compete with the energy sources already 100% functional as the solar energy in Ecuador that reaches levels of nominal generation power during 2018 of 27.63 MW with a percentage of 0.52% of plants with renewable energies.

2. Implemented system

To determine an evolutionary cycle methodology by maintaining phases and evaluating the results obtained with each species, determining the production of electricity in the order of (mv) in order to create the ideal prototype or fuel cell as the experimentation progresses [11].

2.1. Definition of the objective.

The scope of this work is defined as the production of energy through the fuel cell prototype, in a controlled soil environment, generating an analysis of its electricity production with potential plant species in the region.

2.2. Prototype functionality.
The biofuel cells present varieties in their structure and efficiency of electrical energy, applications that generate kilowatt-hours according to the amount of units installed in previous studies of P-MFC [6] our prototype includes a smaller and individual generation for each species of plant, obtaining voltage levels, clarifying that the study leaves the possibility open for an application with scopes where the plants are located in large areas obtaining optimal energy levels.

2.3. System description.
The prototype comprises a series of phases. The first phase consists of the elaboration of the stainless steel collecting mesh made up of sections joined in a parallel manner [12]measuring 100 cm x 50 cm each, a 30 cm copper wire weave was implemented on the mesh (Anode) and then a 30 cm copper wire spiral was made (Cathode), which was integrated inside a plastic container of 100x50x50 cm where several layers of Gravel, prepared soil (layer free of residues), Yellow soil ,activated carbon and finally the plant were placed.
The carbon layer as a component of the biofuel cell medium is essential, affecting the rapid decomposition of nutrients such as phosphorus, nitrogen, potassium, etc., its use was gradual, all components were linked making sure that there are no effects of the use of metals and their interconnection in a medium produce results that deviate from the feasibility of research, we checked the measurement of voltage and amperage with a digital multimeter (Extech EX430A) for temperature and humidity of the environment we used a digital thermohygrometer (HTC-2) was carried out over 30 days with three measurement intervals morning (09:00 am), afternoon (14:00 pm), night (19:00 pm)
The beginning of the second phase is to recognize which are the most suitable plants for the application. Five different types were identified: **Chlorophytum Comosum** (Love Loop-Ribbon), **Capsicum** (chili), **Ceroxylon Alpinum** (Areca Palm), **Codiaeum Variegatun** (Croto Amarillo) and **Aloe Vera** (Sábila), all were selected for their resistance to the climate of the coastal region of Ecuador and the average need for water.

3. Results
The results of the application of the fuel cell are shown, of the 5 species evaluated there is a greater energy yield by the species **Ceroxylon Alpinum** (Palma Areca), with a value of 0.81 v, the trend of yield for the other species of plants is presented in a continuous way in a range that does not vary during the days of sample.

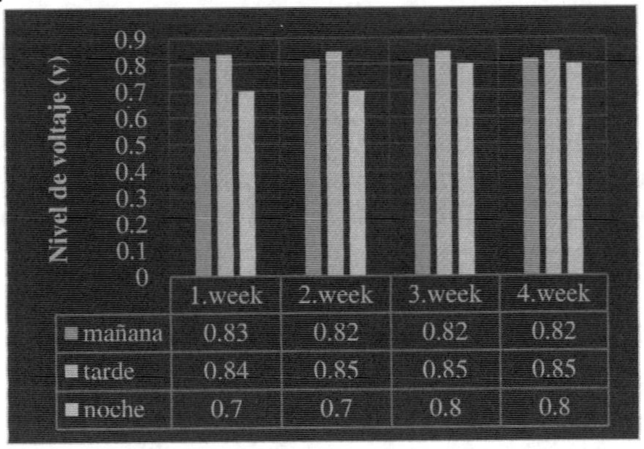

Figure 2. Species energy trend Ceroxylon Alpinum (Palma Areca)

3.1. Ceroxylon Alpinum (Areca Palm)
In figure 2, the energetic trend of the plant is observed, plant of tropical regions, with an adaptability to warm climates and environments with little incidence of water, of medium size the prototype of fuel cell throughout 30 days with an average environmental temperature of 26- 27° centigrade, obtained an average of high generation 0.806 volts.

IOP Conf. Series: Earth and Environmental Science **633** (2021) 012006 doi:10.1088/1755-1315/633/1/012006

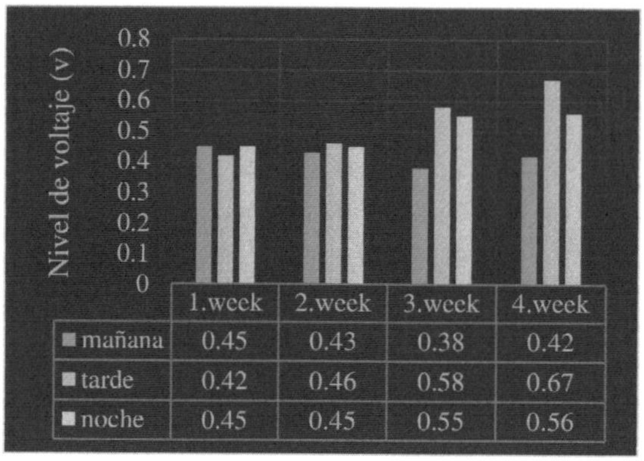

Figure 3. Species energy trend Chlorophytum Comosum (Lazo de Amor).

3.2. Chlorophytum Comosum (Love Loop-Ribbon).

In figure 3, the energetic trend of the plant is observed, plant from tropical regions, with adaptability to hot and cold climates with low incidence of light, small size fuel cell prototype over 30 days with an average temperature of 26 °- 27 ° C obtained a high average generation 0.485 volts.

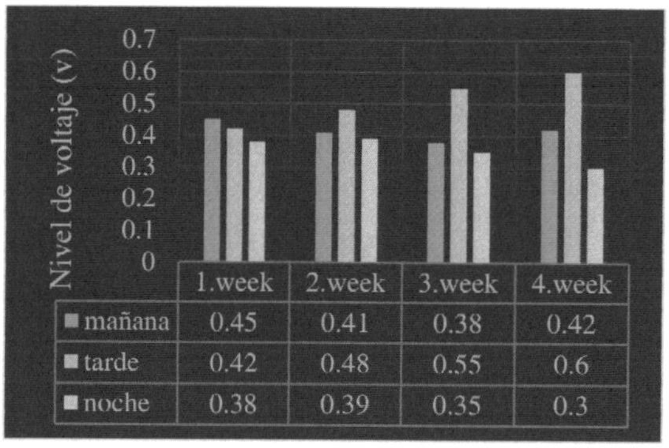

Figure 4. Species energy trend Codiaeum Variegatun (Croto Amarillo)

3.3. Codiaeum Variegatun (Croto Amarillo).

In figure 4, the energetic trend of the plant is observed, plant from tropical regions, with adaptability to hot climates and environments with high incidence of light, of an average size the fuel cell prototype over 30 days with an average temperature of 26°- 27° Celsius obtained an average generation with 0.427 volts.

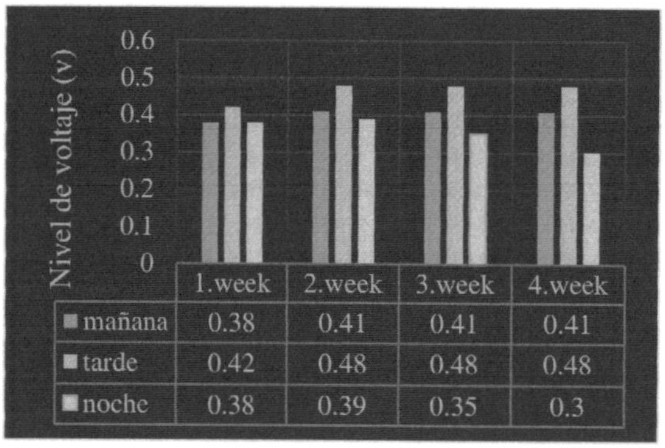

Figure 5. Species energy trend Capsicum (Chili)

3.4. Capsicum (Chili).

In figure 5, the energetic trend of the plant is observed, plant from tropical regions with adaptability to cold climates and environments with high incidence of light, of a medium size, the fuel cell prototype over 30 days with an average temperature of 26 °-27 ° C obtained a high average generation 0.407 volts.

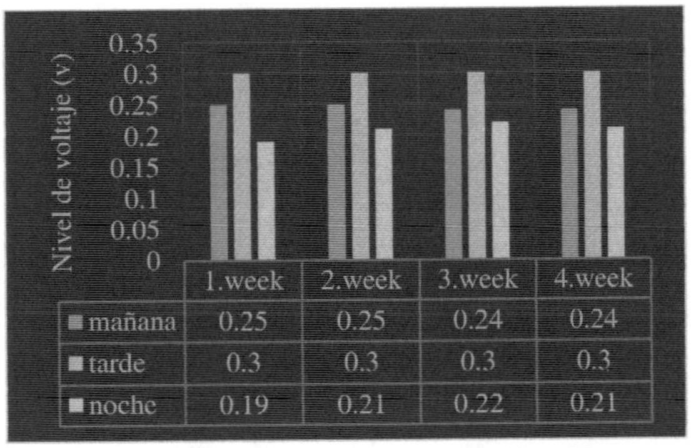

Figure 6. Species energy trend Aloe Vera.

3.5. Aloe Vera.

In figure 6, the energetic trend of the plant is observed, plant from tropical regions with adaptability to hot climates and almost desert environments, without the need for abundant water, with high incidence of light, of a medium size, the fuel cell prototype over 30 days with an average temperature of 26 °-27 ° C obtained a high average generation 025 volts.

The electrical performance of each circuit is within the expected range of 0.8 volts, the conductivity of the anode and cathode was tested, the result of controlling the composition of the soil, contributes adequately to the process of decomposition of organic matter produced by photosynthesis, the species evaluated have differences in their composition, structure, shape, adaptability to temperature, development, however there is an electrical potential in development, within the project factors impossible to control as the climate has been very useful to mark its feasibility of a future prototype.

In the region and with species that have excelled in electrical performance Ceroxylon Alpinum (Areca Palm) reached a value of 0.81 volts, also developed a set of additional tests consisting of an interconnection of several circuits with different species.

Circuit A is made up of 4 species of plants with connection **Codiaeum Variegatun** (Croto Amarillo), configuration **Washingtonian Robusta** series (Fan Palm) new species **Chlorophytum Comosum** (Love Loop - Ribbon) and **Ceroxylon Alpinum** (Areca Palm) where a value of 0.75 volts was obtained, figure 7, refers to the connection of the series circuit of the plants.

Figure 7. Open circuit voltage, series connection.

The serial test was conducted as the third phase of the project in order to obtain accurate and optimal results, showing in figure 8 the energetic trend established for that series circuit ,the value obtained is close to reaching the average range of 0.80 volts does not exceed the index; however, the analysis shows that the functionality of the fuel cell is efficient, the determining factor would be the variety of species

It is necessary to point out that the production of organic matter that decomposes to promote the oxidation and later production of electrons and proton components is much more complex and outdated within the same circuit.

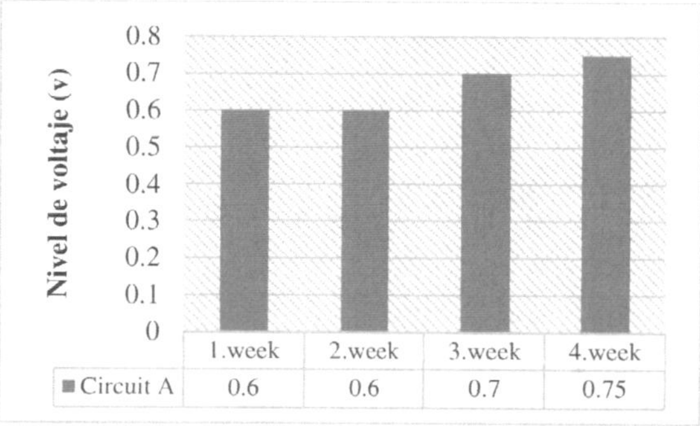

Figure 8. Energy trend, circuit configuration in series analysis with 4 different species for 30 days

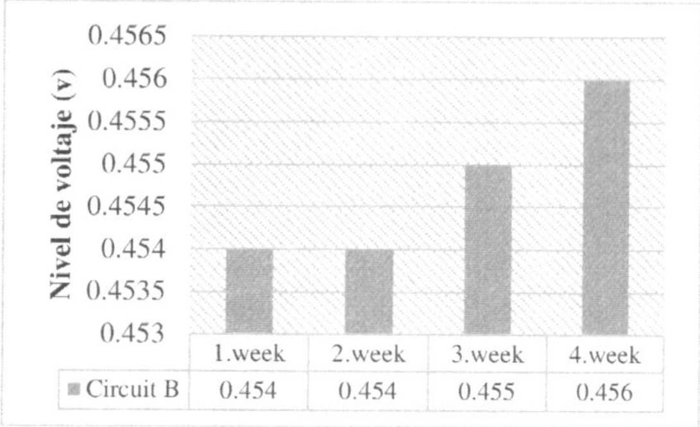

Figure 9. Energy trend, circuit configuration in series analysis with 3 different species for 30 days.

Circuit B determines the results, reaching a maximum of 0.456 volts, during the 30 days of measurement, the species interconnected between each other maintain the initial characteristics, as in the example, figure 9 refers to the energetic trend produced in the interconnection of the species.
Before the incidence of species with little similarity would be the answer for the performance in terms of electricity production, the species used in this second example of application are
Codiaeum Variegatun (Croto Amarillo), a species already evaluated, **Washingtonia Robusta** (Fan Palm) and finally **Chlorophytum Comosum** (Love Loop - Ribbon), the electricity produced by the decomposition does not exceed the voltage range of 0.8 volts in this configuration applied.

4. Discussion

The conventional electricity supply in the region is prioritized by the use of fossil resources and represents a technology with years of application, achieving a position at the top of the economies of electricity production, our prototype fuel cell integrating a photosynthetic process can be improved considerably from the point of view of structural design so that the cultivation of large systems linked together is possible.
If the fuel cell manages to position itself as an efficient generation technology in the coming years, it can be assumed that it will similarly gain an economy of scale advantage. This will influence the environmental design comparison between fossil-based electricity supply and fuel cells, we have established the limitations of the system related to a minimum scale of study and application of current location and time space, the electricity yield is consistently based on a given region of the Ecuadorian coast, unlike current energies according to [13] which no longer maintain this limiting factor, the key to success lies in the systematic evolution of the project maintaining energy sustainability and the use of indigenous resources that biotechnology offers.
The synchronism between the future competitiveness of these options makes possible the technological development in the electrical markets at world level.

Ecuador is characterized by its close relationship between the exploitation of resources and their preservation in equitable ways, the electrical area is developing gradually in relation to other countries where the use of fossil fuels has decreased dramatically, and new technologies are an important part of their internal electrical markets producing, and commercializing at small and large scales [9]

5. Conclusions

It is possible to conclude that biofuel cells, capable of integrating photosynthetic processes to their principle of operation, reveal the potential to become the new technology producing electricity with great scope after its development, reaching self-sustainability and overcoming the problem of fossil fuels and all collateral damage of use today, we can define several factors to support this conclusion one of them is the accessibility of implementation of prototypes, especially in communities where conventional energy does not reach constantly, in Ecuador remote areas lack electrification according to [10] so the lifestyle is limited to the use of diesel-based energy consumption techniques for lamps, long walks from one village to another to access telecommunications service, refrigeration of food, the use of lithium batteries for flashlights ,radios or other devices that do not have an electrical power supply is common ,pollution is present with their consumption.

Another factor that makes our study viable and supports the theory of application as a new technology of electrical production is its design at low cost, in figure 10 we can see the operation and descriptive design of the applied prototype the materials used for the study are accessible with affordable prices, providing a point for its broad construction of large-scale systems with projections of electricity production in order of Kilowatt hours per square meter, we can also cite as a factor pressing their various types of applications, throughout our writing we have seen as similar prototypes in totally different environments as the bacterial cell have been used in small pilot devices by Plant-E (2009) [11] which are marketed with a much wider measuring range and with adaptability to extreme climates reached in European winter seasons

If we land and adapt examples, we can place our prototype for applications in joint lighting systems in recreational spaces, or signalling of short areas of roads, the levels obtained continuously over 30 days with species of **Ceroxylon Alpinum** (Areca Palm) of 0.8 volts crystallize the applications described.

Evaluating in a general way each phase for which our project has been submitted, we conclude that it is not possible to make a real comparison in this timeline with already established sources of energy production, the scale of differences is significant in important aspects from the magnitudes of generation, to scopes of systems, we have made clear the existing potential of the prototype during the whole process, projections point to its gradual evolution in electricity generation and scope, materializing the initial objective, the generation of electricity through fuel cells with research and development is key to a country of diversity.

Figure 10. Operation diagram, fuel cell

6. References

[1] D. d. E. E. Y. e. al, ARCONEL. Atlas of the Ecuadorian Electricity Sector 2017, Quito-Euador, September 2018.

[2] C. R.-L. a. F. J. J. Lata-García, «Attaining the Energy Sustainablility: Analysis of the Ecuadorian Strategy Towards Energy Sustainablility: Analysis of the Ecuadorian Strategy,» Vols. %1 de %213, no. 1, pp. 21-29, 2018.

[3] E. E.-. M. a. S. F. M. E. EL-Khouly, «Solar energy conversion: from natural to artificial photosynthesisi,» *Journal of photochemistry and photobiology C: Photochemistry Reviews,* vol. 31Elsevier B.V, pp. 36-83, 01-Jun-2017.

[4] M. H. e. al, «Electricity production with living plants on a green roof: environmental performance of the plant-microbial fuel cell,» *Biofuels, Bioprod Biorefining,* vol. 7, nº 1, pp. 52-64, Jun 2013.

[5] H. Q. G. a. R. Q. L. Z. G. Li, «Electrical stimulation boots seed germination, seedling growth and thermotolerance improvement in maize (Zea mays L),» *Plant Signal. Behay,* vol. 14, nº 12, Dec. 2019.

[6] A. M. G. A. M. M. a. T. H. T. A Khattab, «Luminescent plant root: A step toward electricity-free natural lighting plants,» *J. Mol. Struct,* vol. 1176, pp. 249-253, Jan 2019.

[7] S. S. T. O. A. K. N. N. a. K. R. F. Bantis, «Current status and recent achievements in the field of horticulture with the use of ligth- emitting diodes (LEDs),» *Sci. Hortic, (Amsterdam),* vol. 235, pp. 437-451, May 2018.

[8] S. Z. a. L. T. D. O. Ardavani, «Redesigning the exterior lightinh as part of the urban landscape: The role of transgenic bioluminescent plants in mediterranean urban and suburban ligthing environments,» *J. Clean. Prod,* vol. 242, Jan. 2020.

[9] R. P. S. a. M. P. S. A. K. Akella, «Social economical and environmental impacts of renewable energy systems,» *Renew. Energy,* vol. 342, pp. 390- 396Feb. 2009.

[10] E. M. M. H. S. A. Z. Silva, FACULTY OF ENGINEERING.

[11] R. Timmers, Electricity generation by living plants in a plant microbial fuel cell.

[12] B. C. H. S. a. A. Heinzel, «Materials for fuelcell tecchnologies,» *Nature Publishing Group,* vol. 414, nº 6861, pp. 345-352, 15-Nov-2001.

[13] «Generating current from plants plant-e technology,» [En línea]. Available: //www.irdindia.in/journal_ijraet/pdf/vol3_iss10/. [Último acceso: 30 May 2020].

IOP Conf. Series: Earth and Environmental Science **633** (2021) 012007 doi:10.1088/1755-1315/633/1/012007

A review on Hybrid Electric Vehicle and simulation on Hybrid Electric Vehicle Drivetrain

Yagnesh Bapodra[1] and Udayakumar Rajamanickam[2]

1 BITS Pilani, Dubai Campus, Dubai, UAE.
2 BITS Pilani, Dubai Campus, Dubai, UAE.
E-mail: bapodrayagnesh@gmail.com

Abstract. This paper discusses about the Hybrid Electric Vehicle (HEV) and the simulation on the Parallel Hybrid Electric Vehicle Drivetrain. The concept and the classification of the HEV is discussed here and the simulation and analysis of a drivetrain for Parallel Electric Hybrid Engine (PHEV) was formulated on MATLAB Simulink. The methodology of simulation was to accelerate or decelerate the vehicle independently with the electric motor with engine running at constant speed. And the parameters such as electric motor and engine power were compared and observed. And research on CO2 emission and fuel consumption was made between the PHEV and Conventional Vehicle (ICE Vehicle). And the conclusion was made according to the research and the simulation that the development of HEV will have a great potential in fuel economy improvement.

1. Introduction

This paper discusses about the Hybrid electric vehicle and the simulation on the Parallel Hybrid Electric Vehicle Drivetrain. The first Hybrid Electric Vehicle (HEV) was invented in 1889 by Willian H. Patton and was used in trams and small locomotives and later it was used smaller vehicles such as cars, SUVs, etc but it were not successful because of no oil shortage at that time but the demand grew up in late 1900s during the rise in oil crisis and then many companies shifted their agenda from performance vehicle to efficiency and fuel economical vehicles Such as Toyota Prius which was introduced at that time was having great impact on the market and nearly in 2008, the were many HEVs in the markets and with better efficiency. The EHVs were a booming concept, so every company launched this kind of vehicle like Range Rover Hybrid, Honda Vezel-Hybrid SUV, Audi Q5-Hybrid, BMW 5-Series-Active-Hybrid, BMW 3-series-Hybrid, Ford C-Max-Hybrid, Acura ILX-Hybrid, etc.

A Hybrid Electric Vehicle (HEV) is a type of hybrid vehicle which consists of combination of two power source to drive the vehicle. The two power sources are ICE (Internal Combustion Engine) and Electric Propulsion System (electric source of energy). HEVs basically works efficiently by switching the power source of the vehicle from ICE to Electric propulsion or vice versa as per the driving conditions or can be driven using both the power source on high performance conditions of the vehicle. Basically, the aim of these concept was to build a fuel efficient and less emission car power source. The concept has been a successful and it has been used in many vehicles including trucks, buses and pickup trucks.

The Drivetrain is known to be a group of components which transfers the power generated from the engine to the driving wheels excluding the engine or power source and the wheels. And a simple drivetrain transfers the power generated from the engine to the transmission which is engaged with the flywheel by clutch and the power from the transmission is transferred to the rear differential by long propeller or drive shaft from where power is transferred to axle and from axle to the two wheels which drives the vehicle. There is also a term called 'Powertrain', the difference between the power train and

drivetrain is just that the powertrain includes all the drivetrain components and includes the power source components such as engine and wheels.
The Drivetrain changes depending on the type of the vehicle such as Front-wheel, Rear-wheel or Four-wheel drive etc. or on the type of the power source such as for electric vehicles, hybrid electric vehicle, Hydraulic hybrid vehicles etc. The hybrid electric vehicle Powertrain has more no of components and more complexity than a simple ICE Powertrain due more electrification. In HEV powertrain, electrical energy is also used as along with engine to produce torque, so the drive train are designed in such a way that the torque of both sources is combined and distributed to the wheels. And the drivetrain uses the power control units to control or stop the power from the engine or electric motor.
Basically, the HEV drivetrain consists of ICE, Electric motor, Battery, energy convertor, Transmission and Power Control Unit (PCU). The HEV drives their wheels on two power source ICE and Electric Motor which are connected to Transmission which are connected to the axle and then to the wheels. The Electric Motor is powered by battery, it is connected to the DC-DC convertor to converts the voltage of the current flowing through it and supply it to the motor. The power sources are controlled by power management unit which commands the sources to switch On/Off according to the performance required and optimization of the vehicle.
The HEV has Power Management unit or Power Control Unit which is built with control strategy **[1], [2]** for power optimization and it controls and commands the power source components based on the driving condition. the control unit continuously observe the vehicle performance parameters and efficiency parameters and to optimize these parameters it commands the power sources of the vehicle. And the control strategy are designed on different conditions such as 1) when the vehicle is travelling with low speed or less acceleration and the performance required is less and the battery is fully charged, so the vehicle operates on Electric motor powered by battery and engine will be shutoff, 2) when the vehicle is accelerating or requires high performance with engine on and the battery charge is full that time, electric motor is started to assist the engine for acceleration and to reduce fuel emissions because at high performance more fuel is consumed, 3) when the vehicle is cruising with engine power and battery charge is low, the engine power is transferred to wheel as well as to the electric motor which becomes generator and generate electricity which is stored in the battery for later requirements, 4) The battery is also charged by other components of the vehicle such as regenerative braking system (RBS) and regenerative shock absorbers (RSA), the RBS supplies electricity to battery during the braking of the vehicle, the energy wasted in braking is used by the generator to generate and store electricity in the battery and similarly the RSA, due uneven roads the vehicle gets impulsive shocks and this shocks are used to generate electricity. These regenerative energy sources decrease the fuel consumption which decreases the emissions and increases the range of the vehicle and increases the economy and efficiency of the vehicle.

2. Classification of HEV

There are three main types of HEV drivetrains which are Parallel mode, Series mode and Power-split: -
Parallel: -The basic HEV powertrain consists of Electric motor, ICE, Generator, battery, power convertor and transmission. In the Parallel HEV the electric motor system and the ICE are arranged parallelly and both combines their power and drives the wheels. And this configuration is more efficient than the series HEV. It can also give instantaneous power during high performance requirements by electric motor assisting the ICE to build more power. But it requires a complex linkage between the engine and motor shaft to join and give combined power.
Series: - In the series HEV the engine doesn't have any mechanical linkage with the wheels, and they don't have any transmission because the wheels are driven only by the motors. The engine is connected to the generator which generates the electricity then with help of charger it is stored into the battery, then from the battery the controllers supplies the current to the Electric Motor and which drives the wheels. the Series HEV can run on motor independently when the engine is Shut Off. The efficiency of the vehicle is less than the Parallel HEV due more no of components arranged in series HEV which increases the energy losses.
Power-Split: - Basically it is combination of both the Series and Parallel HEV powertrain, the Power-split HEV power train starts from the engine which produces the power from where it is divided in to

IOP Conf. Series: Earth and Environmental Science **633** (2021) 012007 doi:10.1088/1755-1315/633/1/012007

two paths where one goes directly to the transmission and drive the wheels like conventional vehicle whereas other goes is connected to the generator which generates electricity and stores in the battery and then later use it to drive the motor which is connected after the transmission directly to the driving shaft. So here both Series and Parallel concept is used for getting the best efficiency of the vehicle according to the driving conditions.

There are also other types of HEV on basis of power source mainly electric power source such as Full HEV, Mild HEV, Plug-in HEV, Micro-hybrid etc.

Full HEV: - This kind of EHV can have independent power source like working just on electric propulsion system or ICE or both can work at same time to power the vehicle during high vehicle performance requirement and it electric battery doesn't required to be charged from external charging plug but the battery gets charged by ICE itself so no external battery charging is required. They can also operate in parallel and series mode also example like "Toyota Prius". The parallel mode uses both the electric motor and combustion engine at same time to drive the wheels. whereas the series mode means the main power is the ICE and the electric motor will assist the ICE for more performance.

Mild EHV: - This kind of EHV also has both power sources but here the vehicle cannot work independently on the electric propulsion, here the vehicle mainly runs by both source of power at same time, but it can work at ICE independently. Example: - "Honda Accord Hybrid". they are limited to parallel mode, so they work only on series mode.

Plug-in EHV:- This kind of EHV also has two power source and can run independently on electric or ICE but here the battery of electric propulsion system is not charged with the ICE but it is charged by the external charging plug that's why it is named Plug-in EHV. Mainly the Plug-in hybrid have more capacity in battery then in Full hybrid and they range is also higher than the average Full hybrid. They can work on both series and parallel mode.

There are some other types of EHVs which are not common in use like high-performance hybrids which are used in high performance cars such as "Ferrari", "McLaren", Etc. These Automobiles companies uses this hybrid technology to boost cars performance unlike others they do not focus fuel efficiency and they focus on performance maximization. They use the hybrid electric motors to assist the engine to maximize performance to the vehicle they use the hybrid electric technology in completely opposite way.

The HEVs are fuel efficient than conventional vehicle because of less emission generate due to electric motors. And they are cost-effective and affordable also, compared to conventional vehicle. The ICE required in HEV is lighter than the conventional one, so it makes the Vehicle lighter and more spacious. the Vehicle also decreases fuel usage which increases the driving range of the vehicle. The HEVs are also more preferred than pure Electric Vehicle (EV) [3] due to performance compromise in EV, EVs are costly and requires electric charging stations for charging which are not available outside the cities, obviously lesser emission are there but the time required for charging is 10-12 hours compared fuel engine which is too much. but in HEV, charging stations are not required for charging the battery, but it is charged by the generator, which is connected to the engine, so the power gets reused by electric motors to accelerate the vehicle. The HEVs mainly are built for economy, so there is a compromise of performance in some HEVs, the HEV also increases the maintenance of the vehicle because of complex components used and mainly due to electronic components requires skilled experts for the maintenance.

3. Simulation methodology

The objective here is to simulate the drivetrain for Electric Hybrid Vehicle (HEV) and study and analyze the different parameters associated with the drivetrain and its components. The simulation of powertrain is done for parallel type of HEV, in the software called MATLAB Simulink. The Parallel HEV consists of an ICE, DC-DC convertor, Battery, Electric Motor, transmission and two wheels (Because half-Vehicle model is built, and only rear wheels are included).In the PHEV the ICE and Motor

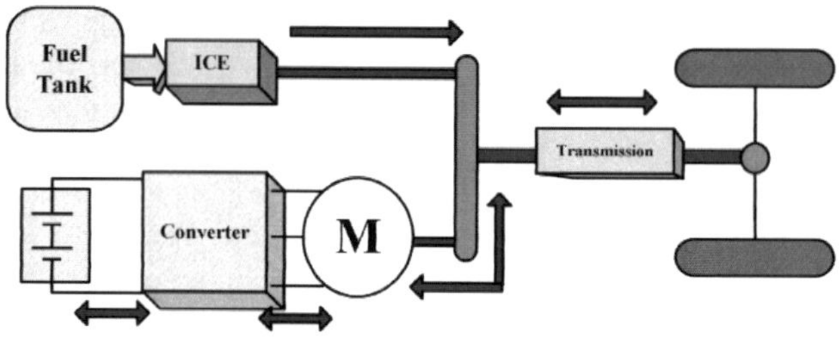

Figure 1. Parallel HEV diagram

output shaft are connected with mechanical coupling or any other linkage mechanism to combine and transfer the power to the transmission.

In the simulation the engine and motor are arranged parallel to each other and unlike the figure 1 above only engine is attached to the transmission and the battery is attached with the DC-DC convertor which supplies electric current to electric motor to drive the shaft and the output shaft of both the transmission and the motor are combined to combine the power produced by both the power sources. Due lack of skills in MATLAB Simulink the control strategy unit was not developed in the powertrain model to command the power source components according to the performance requirements of the vehicle. Therefore, the method used here for knowing the power requirements of the electric motor to drive the wheels was, the engine was kept On at constant throttle to deliver the power to wheels to maintain an initial speed of the vehicle. And the electric motor is used to accelerate the vehicle and maintain a constant speed and then deaccelerate back to the initial speed which means electric motor was switched On during the performance requirement and maintain the performance and then the electric motor is switched Off and the vehicle is cruising with the engine power supplied to the wheels. The Electric motor is managed with the speed command strategy made with help of MATLAB documentation and the electric motor speed is increase and decreased with this strategy. And by changing the electric motor speed the simulation results or the parameters are obtained such as electric power, shaft speed, battery and fuel Consumptions, speed of the vehicle and the battery losses.

The components used here were an engine which was an ideal spark-ignition engine which has simple parameters, and which has max speed of 7000 RPM and max power is 50 KW. And the Gearbox selected here is a simple gearbox with constant efficiency 95%. The electricity is supplied from the battery where it is stored, and it consists of DC voltage source, internal resistance, current and voltage sensor, the DC voltage source is where the electricity is stored and used, the measure of the battery power is done by the voltage and current sensors. And DC-DC convertor is used is an electric-mechanical device which converts the voltage of DC current from high to low or low to high but here during the power conversion, the voltage is increased because the motor requires more voltage current. and the electric motor is used, it was a block set taken from the MATLAB Simulink library sources, basically this converts the electrical energy to the mechanical rotational energy and vice-versa when the engine powers the motor it changes to the generator and charges the battery. The electric motor and the speed command strategy were made with help of the documentation of MATLAB. The electric motor was commanded with inputs in the speed command strategy to accelerate and deaccelerate accordingly at different step timing. The wheel which were used here were also a simple tire which doesn't have any slip, and which were connected by axle to the main driving shaft coming from the transmission. The Vehicle body dynamics were also measured but assumption was made, the slope of road and the wind speed were not considered and set to zero. the vehicle body was used to measure the normal force acting on the wheels and the speed of the vehicle during different performance.

The Speed command strategy here means to give the command to the electric motor to increase or decrease its speed. So here four inputs are given first is to accelerate motor at a given step time 20s and similarly second input command is given to stop acceleration at 70s to get a constant speed of the vehicle and then the vehicle was deaccelerated by giving input at 120s and giving command to stop acceleration at 170s and then the vehicle was running only on the engine. And these all input signal are combined to one signal by integrating them then they are converted to RPM and it are measured by sensors connected

IOP Conf. Series: Earth and Environmental Science **633** (2021) 012007 doi:10.1088/1755-1315/633/1/012007

to it for result purpose. And the total duration of the test was 200s and all the parameters such as engine power, battery power, electric power, speed of engine and the electric motor, fuel consumption, etc. were determined by different sensors connected to each component of the simulation.

According to some research [3],[4],[5], the comparison between the Parallel Hybrid electric vehicle (PHEV) and the conventional ICE Vehicles (ICEV) were made by collecting the data for every parameter in real world driving of both the vehicle type PHEV and ICEV and compare them. The Software use here was known as Autonomie. the simulation was done on several driving conditions and many real-life parameters were also measured such as weather conditions, accessories load, road pattern, vehicle activity (driving cycle), and engine capability and also fuel economy, emissions, and all equipment performance were evaluated. The standard driving cycles which were used are FTP, UDDS, and US06HWY. According to results, the fuel economy of PHEV was increased up to 68% in real driving condition than the ICEV the reason was due to regenerative braking which was used continuously in city due to which the CO_2 and other harmful gas emission decreased up to 40% due using electric mode during less performance required by the vehicle generally in the city traffic. Another benefit was for ECU which can change the power source unit according to required performance, which lead to increase in engine performance by 12% in real world driving cycle whereas in FTP it was 30%.

4. Simulation results

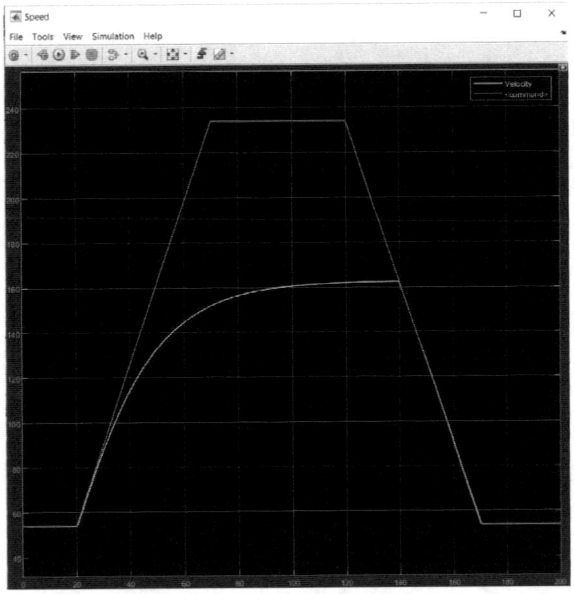

Figure 2. Vehicle speed Vs Motor speed

The simulation results we get at 200s step time are the speed of the vehicle and the motor speed as in figure 2. The motor speed is converted from rpm to kmph for comparing it with the vehicle speed. The speed of motor is increasing at 20s constantly with time. The motor powers the wheels of vehicle body that why the speed of the vehicle is also increased gradually with respect to the motor speed. At 70s to 140s the motor has stop accelerating and the vehicle speed is also getting constant, at the 120s the motor decreases its speed to same as the initial speed and the vehicle speed also gets deaccelerated.

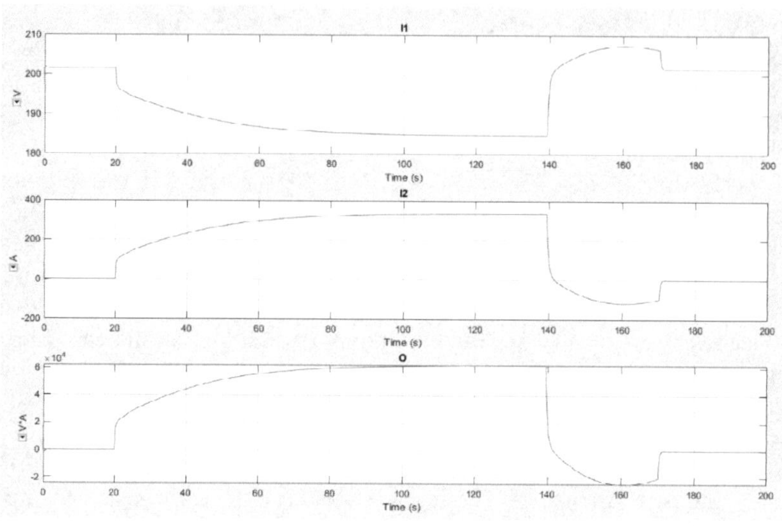

Figure 3. A) Voltage B) Current C) Electrical Power

The figure 3 above represents the voltage, current and the electric power graph versus time plot. The voltage and current are calculated by ammeter and voltmeter at every time interval, present in battery circuit. The Electric power is measured by product the voltage and current input value, we get the power curve. And as in the graph the power increases at 20s due to acceleration by motor to 20 KW, then motor attends the stable speed so it consume power up to 60 KW and decreases at 140s due to deacceleration and then it shows negative power that means the engine shaft power is also transferred to the motor/generator to recharge its battery and after recharge it gains back 0 KW.

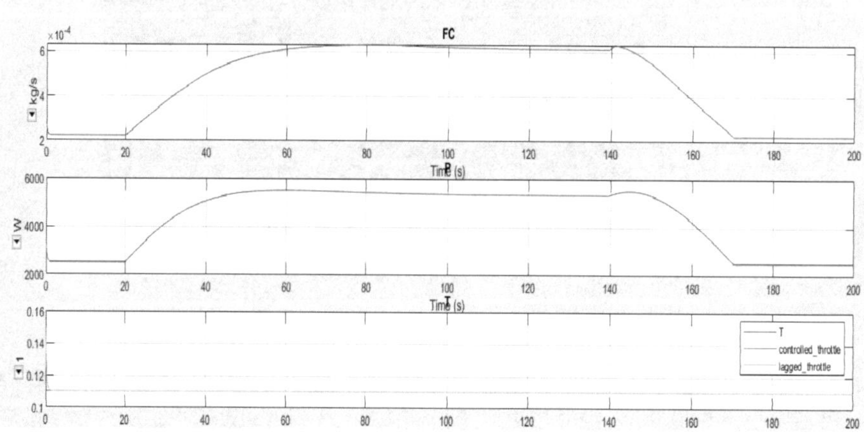

Figure 4. Engine Parameters A) Fuel consumption B) Power C) Throttle

The figure 4 above shows the engine parameters with respect to the time the A) is the fuel consumption as u can see the fuel consumption is increased at 20s up to 0.0006 Kg/s because the vehicle wheels are powered by motor so there is less fuel required than actual amount to accelerate the vehicle by the engine only and then decreases again back to around 0.0002 Kg/s for maintain the vehicle speed. B) is the Power of the engine which is also increased during the acceleration is slightly increased due to some increase of speed of the shaft. And after 140s when the motor stops powering the drive shaft, the power required by engine is more to charge the battery by supplying rotational power to the motor and the wheels. C) is the throttle graph which is made constant value 0.11.

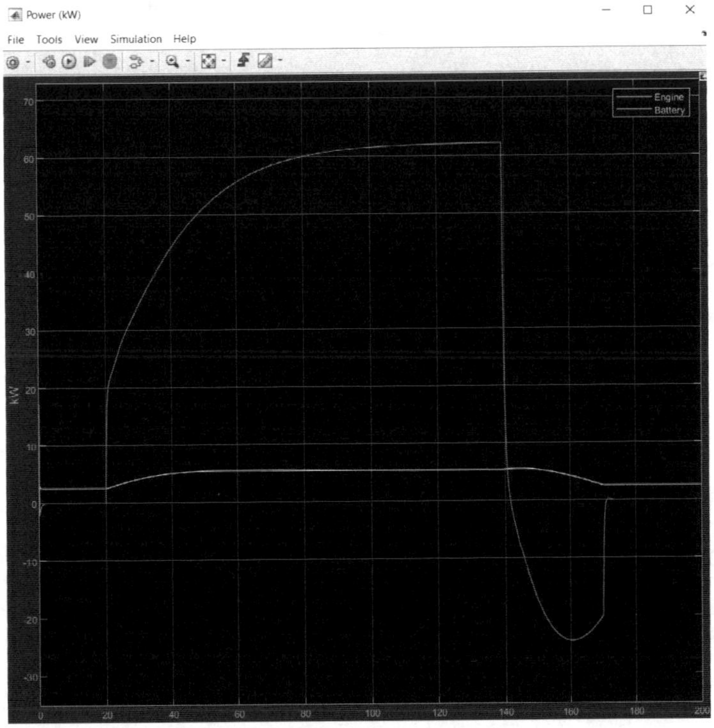

Figure 5. Power Curve of Engine and Battery (Motor)

The Figure 5 above shows battery/motor and engine power Vs time, as we know for acceleration the electric motor is only used and engine is to maintain the vehicle speed, so when the acceleration starts the motor starts powering at 20s so it as in figure the motor power is increasing and the engine power has slight changes. The motor decreases after the 140s second because vehicle starts to deaccelerate. Negative curve after the motor stops, in the figure it shows the power that is absorbed by the motor to charge some amount of battery from the rotating engine shaft and some amount of power is gained due to regenerative braking.

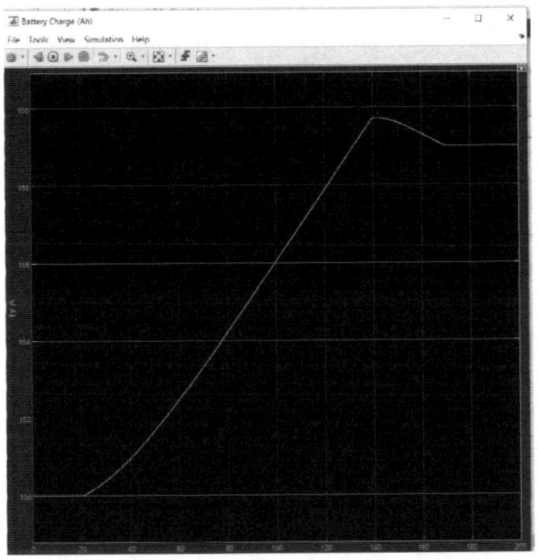

Figure 6. Battery Charge curve

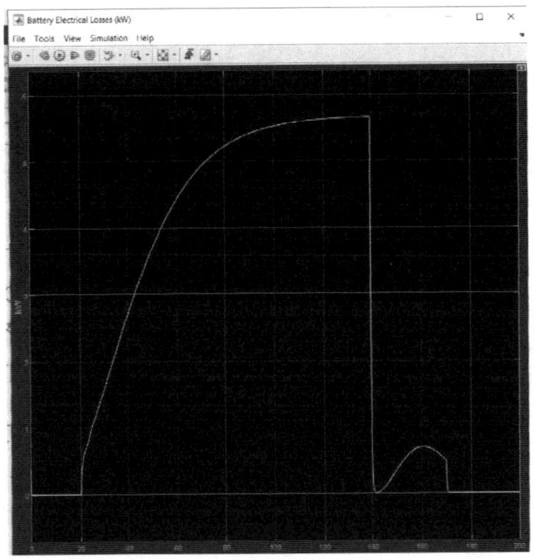

Figure 7. Battery Electrical losses

The figure 6 above shows the battery charge usage and battery losses is shown by the figure 7 above. Figure 6 above shows the battery charge used by the electric motor during the whole process as the acceleration increase the battery charge usage also increases and it used until the deacceleration of the vehicle and then it decreases. The figure 7 above shows battery losses due various reasons, it increases during the acceleration due more current flow and decreases during deacceleration.

5. Conclusion

The simulation of a parallel Hybrid electric powertrain was completed. The method of the simulation was that the engine was powering the wheels to maintain a speed and the vehicle was accelerated and deaccelerate by electric motor, engine was not used to accelerate or deaccelerate. The electric motor was commanded by a speed command strategy and speed controller. The engine was having a constant power due constant throttle. And the power and speed of both engine and motor were determined for 200s step time. The power curves showed that the vehicle has the electrical power which is higher than the engine power and it proves that the vehicle can also accelerate at high speed with only the electric motor. The electric motor also uses the regenerative brake to absorbs some of the power to recharge the battery. And the charging level of battery was also measured with respect to time it also shows that the battery. The charge level also decreases with the time due to increase in motor power and the it also gains some power during the deacceleration due regenerative braking.

The vehicle velocity also increases and decreases as per the figures discussed above with respect to the electric motor acceleration and deacceleration respectively, which shows the output of the simulation as speed of the vehicle. The simulation shows that the vehicle can run on electric motor independently for performance requirement in parallel mode of HEV powertrain. According to studied research [3],[4],[5], it can be concluded that PHEV improves the fuel economy as well as engine efficiency in standard driving cycle as well as real world driving cycle. The improvement in engine efficiency also increased the life span of the engine. Therefore, from all the about research and simulation done it can be stated that the PHEV or HEV has very promising potential solution for improvement of fuel economy than the ICEV conventional vehicles.

6. References

[1] Optimization-Based Control Strategy for Large Hybrid Electric Vehicles-Jianhua Zhou, Mian Li, and Min Xu Shanghai Jiao Tong University, china ,2018.

[2] Energy Management in Plug-in Hybrid Electric Vehicles: Recent Progress and a Connected Vehicles Perspective - Clara Marina Martínez, Xiaosong Hu*, Dongpu Cao*, Efstathios Velenis, Bo Gao and Matthias Wellers, IEEE, 2016.

[3] A Comparison of Electric Vehicles and Conventional Automobiles: Costs and Quality Perspective, Marek Palinski, 2017.

[4] Study of emissions and fuel economy for parallel hybrid versus conventional vehicles on real world and standard driving cycles, Ahmed Al-Samari, Mechanical and Aerospace Engineering Department Alumni, West Virginia University, 26505, USA, 2017.

[5] How hybrid-electric vehicles are different from conventional vehicles: the effect of weight and power on fuel consumption, C Reynolds and M Kandlikar, 2007.

IOP Conf. Series: Earth and Environmental Science **633** (2021) 012008 doi:10.1088/1755-1315/633/1/012008

Development of X-in-the-Loop testing system for electric vehicles

K Karpukhin[1]

1Project Director, NAMI Russian State Scientific Research Center, 2 Automotornaya st., Moscow, Russia, 125438
E-mail: K.Karpukhin@nami.ru

Abstract. The paper describes an effort towards development of electric vehicles in the form of an international collaboration aimed to create an environment that combines laboratory facilities and virtual models into a single research and development system called X-in-the-Loop. The facilities included in that system are connected and shared between developers and researchers working in different geographic locations with different vehicle subsystems allowing them to synchronize their experiments in real time studying the entire vehicle rather than its separate components. The main section of the paper focuses on the part of the X-in-the-Loop project dedicated to the thermal management system (TMS) for high voltage components of the considered electric powertrain. In particular, descriptions are given regarding the control principles and the system architecture used to integrate the TMS into the X-in-the-Loop environment. The virtual part of the TMS is verified by means of simulations. Finally, the design of the thermal management system embedded into the test installation is presented in the form of a CAD model.

1. Introduction

Testing methods known under the common name of "X-in-the-loop" (where "X" can stand for either a hardware system or its separate components) constitute a synthesis of physical tests and simulations [1-5]. These methods imply that a part of the studied object or system has a hardware form while the remaining part is virtual. The operating environment can also be either physical or virtual or a combination of these. Bilateral interactions are established between the mentioned parts in the forms of physical actuation, calculations, and data exchange. Operation of the virtual part is real-time synchronized with the operation of the hardware part. These features allow a researcher or a developer to reconstruct the entire studied system within its operating environment and conduct laboratory tests, which are equivalent to those usually performed at specialized proving grounds with carefully adjusted and/or planned ambient conditions. Further development of this technology is to step beyond the limits of a single laboratory and involve several geographically separated testing facilities holding different components of the studied system, either hardware or virtual, to connect these components and make them operate as a single system. In this case, the real-time synchronization is implemented via a network connection (e.g. using the Internet) between the controllers and software components of the involved testing facilities.

When geographically extended, the X-in-the-loop technology (hereafter abbreviated as XiL) can be advantageous in studying and developing electric vehicles (EV). It is a state-of-the-art engineering practice that powertrain components of an EV have to be integrated into a coordinated chassis control architecture [6]. Specifically, a traction electric drive is usually integrated in a traction control system, braking system (to provide a regenerative braking feature), and active safety systems such as electronic stability control and torque vectoring control [7]. The coordinated control implies a highly coherent real-time interaction between different chassis' systems and components during the powertrain operation. This interaction is usually elaborated during the powertrain development process in both virtual and physical domains [8-10]. Usually, powertrain components are supplied by several producers and developers. Each of them has its own testing facilities and simulation software optimized specifically for the produced component or system. Often, such facilities are equipped with advanced and unique installations or programs that could be unavailable for other developers. To avoid an unnecessary waste of resources and gain maximum benefits from the available specialized facilities, the latter can be connected and shared within a XiL environment that unites all the developers and researchers involved in the elaboration of an EV. An example of such a XiL environment is currently being developed by a European consortium of developers, producers and researchers [5, 11]. The project encompasses all of the above-mentioned aspects of EV integrated chassis development. Its aim is to modify a production EV by introducing a new advanced powertrain, which is based on four in-wheel electric motors having a coordinated control that allows them to deliver traction and regenerative braking functions as well as functions that improve vehicle's driveability and active safety. The latter is implemented by coordinated operation of the powertrain and active suspensions. The XiL system also includes a driving simulator that provides physical modeling of a passenger compartment allowing human testers to drive the virtual vehicle and estimate its driveability and driving comfort provided by the chassis control system.

There is one more aspect of the EV integrated chassis development that needs to be covered by the described XiL system, namely a thermal management system (TMS) of the powertrain. Principles of the coordinated control also apply to that system, since it has to be sensitive to powertrain operating regimes and react appropriately to changes in energy dissipation of the powertrain's components [12]. The institution represented by the author of this work (NAMI) has a substantial background in developing and studying such systems including those intended for electric powertrains [13, 14]. This has allowed us to join the mentioned consortium and use our expertise to elaborate the corresponding part of the XiL system. The remainder of the article describes the main features of this work including the development of a thermal management system and a testing installation to be incorporated into the XiL environment.

2. Main part

In this work, a concept has been proposed for elaboration of an X-in-the-loop (XiL) environment that provides connection and interaction within a "scattered cluster" of testing facilities located in different countries in order to conduct complex researches of electric vehicles and their components. A distinctive feature of this concept is the real-time synchronization of different testing facilities, which share common models and parameters and form a simulation system that reproduces operation of an EV and its components in real-world driving conditions.

The project's part conducted by NAMI consists in development of an installation intended for XiL tests of thermal management systems for high voltage components of electric vehicles. The installation implements physical simulation of traction electric drive components in respect of their thermal and hydraulic properties. Simulations are conducted in connection with other test facilities hosted by the XiL system.

Analysis of the literature [15-19] shows that, in XiL environments intended for testing of automotive thermal management systems, the physical part constitutes the thermal management system itself while the virtual part simulates the powertrain components to be thermally regulated and their operating conditions including ambient factors. There are two types of interaction between these parts of the system: physical and information. The first type is implemented by a device, which physically simulates thermal and hydraulic behavior of the modeled powertrain component, i.e. the coolant pressure drop and the temperature dynamics that would take place in the actual component having specified operating conditions and input coolant parameters. Information interaction is implemented by feedback and control signals. The former result from sensor measurements conducted by a data acquisition system. The main measured variables are the coolant flow rate, pressure and temperature at the inlet of the physical simulator. These parameters are fed into the model of the powertrain

IOP Conf. Series: Earth and Environmental Science **633** (2021) 012008 doi:10.1088/1755-1315/633/1/012008

component whose output variables are transformed into the command signals for the physical simulator therefore closing the simulation loop.
Models of the powertrain components employed within XiL environments usually constitute lumped parameter systems implemented by ordinary differential equations. They do not impose an excessive computation burden upon the XiL controller allowing for real-time simulations with acceptable accuracy of modeling.
Considering the above principles and the specifics of the designed TMS, the XiL architecture shown in Figure 1 was elaborated.

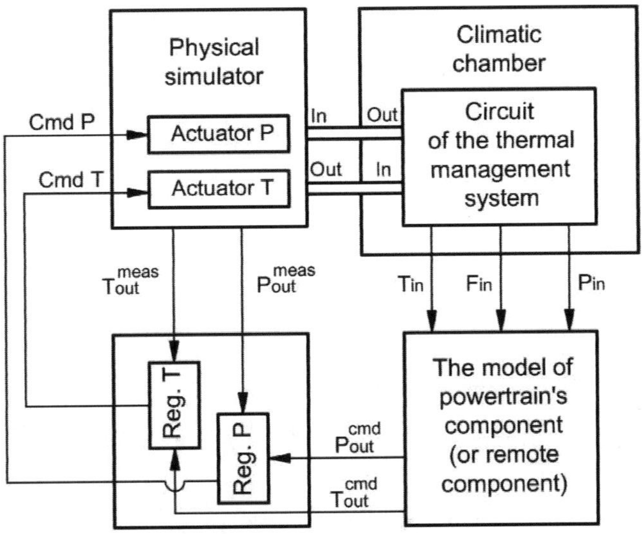

Figure 1. Control architecture of the XiL system

The physical object to be tested is a circuit of the TMS being a functional unit that regulates the temperature of a powertrain component (or a group of components). The TMS circuit is placed within a climatic chamber that simulates the ambient conditions. When necessary, an airflow-making device (i.e. fan) may be installed within the chamber in order to simulate vehicle's motion. The number of simulating circuits (sharing the common climatic chamber) corresponds to the number of powertrain's components.
The physical simulator contains a pipeline whose inlet and outlet are connected to the outlet and the inlet of the TMS circuit. Within that pipeline, there is a device ("Actuator P"), which regulates the hydraulic resistance and, consequently, creates the pressure drop corresponding to that of the simulated powertrain's component. The "Actuator P" is controlled by the "Cmd P" signal. The heating device denoted as "Actuator T", being controlled by the "Cmd T" signal, regulates the coolant temperature.
The virtual part of the XiL system constitutes the model of the electric drive to be thermally regulated. The model calculates mechanical, energy, hydraulic, and thermal parameters of the drive. The main input variables relevant for thermal modeling are the coolant temperature (T_{in}), flow rate (F_{in}), and pressure (P_{in}) at the component's inlet. The corresponding output variables are the coolant temperature (T_{out}) and pressure (P_{out}) in the outlet pipeline. As an alternative (or an addition) to the model, a remote data source containing information on the component's operating regimes can be employed. An example of such a data source is a remote laboratory, which conducts physical tests of the powertrain's component and conveys the measured variables via the internet into the TMS XiL facility.
The controller unit of the XiL system contains the regulators "Reg. T" and "Reg. P", which calculate control signals for the corresponding actuators of the physical simulator. The output variables of the electric drive model are used as the command signals for these regulators. This is highlighted in Figure 1 by superscripts "cmd" assigned to the model outputs. The measured counterparts of these variables

(having "meas" superscripts) are outputted by the physical simulator and used as the feedbacks for the regulators.

Before being embedded into the XiL environment, the models of the electric drive components were tested within a simulation software in order to verify their interaction with the TMS during the powertrain operation. The elaborated mathematical model included the electric vehicle, four wheel-hub motors, and the thermal management system. Figure 2 shows an example of modeling results, which simulate a driving cycle test. The used velocity schedule was the Worldwide Harmonized Light Vehicle Test Cycle (WLTC). The electric vehicle weighed 3150 kg and was equipped with four wheel-hub motors each having 65 kW of continuous power. Operation of the TMS was complicated by a high ambient temperature of 45°C. The following variables are shown on the plots (from top to bottom): vehicle velocity (V), net total power of four electric motors ($4 \times P_{EM}$), total heat (loss) power of four motors ($4 \times P_{EM,heat}$), input and output coolant temperatures in the "warmest" in-wheel motor (T_{EM}). The model was verified in different driving cycles and ambient conditions and was considered suitable for XiL applications.

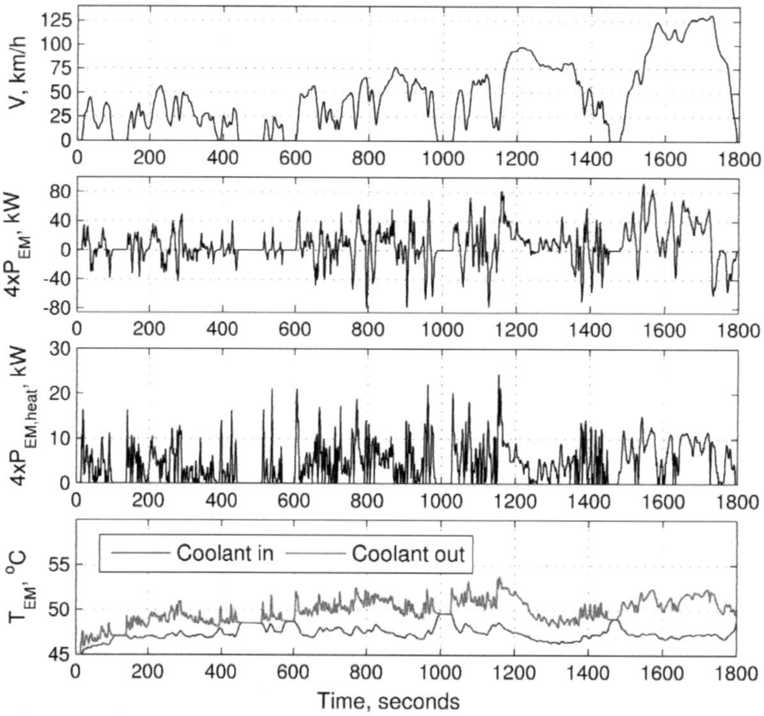

Figure 2. Simulation results for thermal management system operation in a driving cycle

Figure 3 shows the test bench designed in accordance with the elaborated XiL architecture. A distinctive feature of this test system is series-parallel connection between the devices that simulate hydraulic and thermal behavior of the electric motors 4 and inverters 5.

The electric motors and inverters (or their physical models) are arranged into four groups in accordance with the powertrain design that includes four wheel-hub motors. Within each group, the components are connected in parallel and provided with a hydraulic pump. The motors and inverters have dedicated outlets located within the measuring units 8 and intended for disposal of steam-air mixture into the air compartment of the expansion tank 1. The radiator 2 also has a similar disposing outlet located at the top of its casing. The outlets provide regular removing of the air that penetrates into the system's pipelines during its operation or during servicing when the coolant is being changed. To provide feedbacks for the controller of the thermal management system, the measuring units 8 contain temperature and pressure sensors. The three-way valve 9 is intended to warm the system up quickly, when the temperature conditions require so. The electrically driven hydraulic pumps are regulated individually in order to reduce the energy consumption and increase the efficiency of the system.

During a XiL simulation, the system (located at NAMI) receives real-time data from other facilities (situated in the laboratories of the European participants). The data includes the hydraulic and thermal variables outputted by the electric drive components (motors and invertors), as well as information on the vehicle's operation (velocity) and the ambient conditions including the air temperature and humidity. The thermal variables (i.e. the coolant temperatures at the hydraulic outlets of the electric components) are converted from the information domain into the physical domain by means of electric heaters incorporated within the simulating devices 4 and 5. The resulting actual temperatures are measured by the sensors of the TMS and used to regulate operating regimes of the hydraulic pumps and the electric fan belonging to the radiator system 2. At the radiator's air inlet, there is an array of controllable fans intended for simulation of the oncoming airflow corresponding to the vehicle velocity. XiL simulations are to be conducted in a climatic chamber intended to imitate an ambient temperature range from –40°C to +50°C. This is the operating range of the powertrain components specified by the vehicle's manufacturer. In this range, the developed thermal management system should deliver its functions.

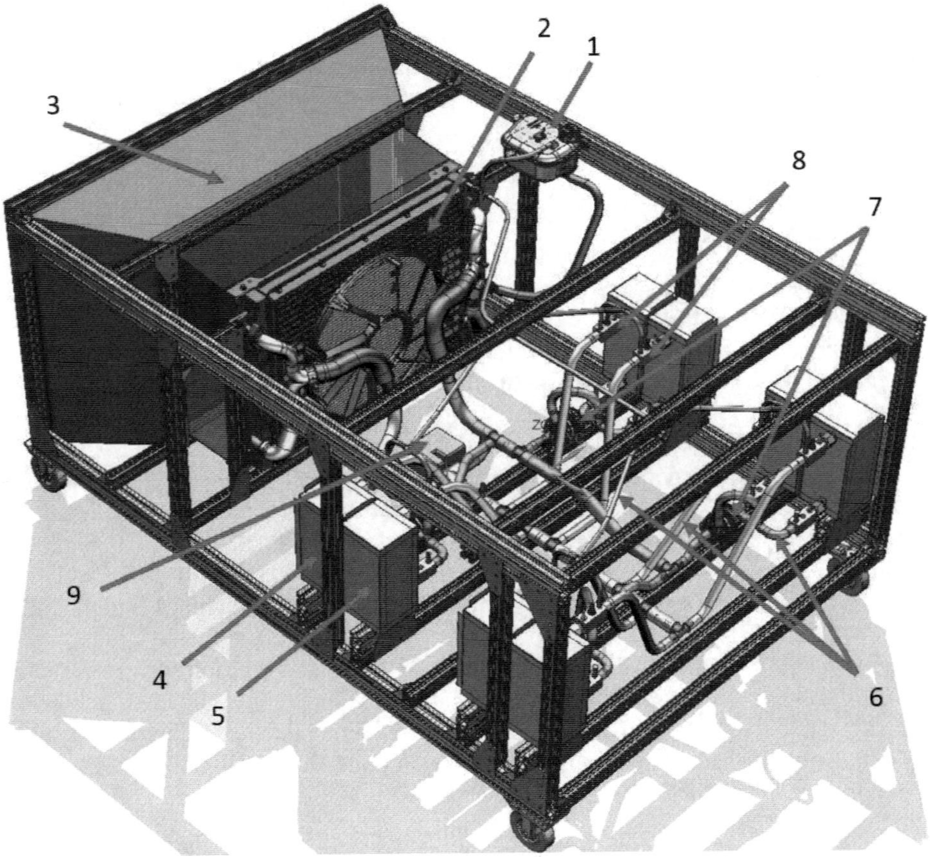

Figure 3. Design of the test bench for the thermal management system of high voltage components of an electric vehicle.
1 – expansion tank; 2 – radiator and fan; 3 – radiator's air inlet; 4 – physical simulator of an electric motor; 5 – physical simulator of an inverter; 6 – pipelines; 7 – electrically driven hydraulic pumps; 8 – measuring units (pressure, temperature); 9 – three-way valve.

3. Conclusions and future work

In the described work, a testing installation has been built in NAMI that simulates operation of the EV thermal management system using a combination of physical and virtual models. The interaction between these models is implemented using a closed-loop control system based on simulations and sensor measurements. The control architecture uses the known principles of building XiL systems and takes into account the features of the considered TMS and the electric powertrain. The powertrain design featuring four traction electric drives (i.e. "in-wheel motor-inverter" pairs) is addressed in both the TMS layout and the testing installation structure. The latter uses a combination of simulating units corresponding to each electric drive and interacting with the TMS through the common model of the powertrain or through data exchange with a remote laboratory. The conducted software simulations allowed for verifying the mathematical models of the vehicle, its powertrain and the TMS including their parts to be used in the XiL environment.

The sequel of this work implies integrating the TMS testing facility into the XiL system and performing experiments in collaboration with the European laboratories of the X-in-the-loop consortium. When conducting simulations, the facility at NAMI will receive real-time data generated at the laboratories of the European participants. In response, the thermal management system will regulate the coolant temperature by means of the electrohydraulic pumps, valves, and radiator's fan control. In turn, the control system of the test bench will collect the thermal variables and feed them back to the remote testing facilities. The experiments are to be conducted in physically simulated ambient conditions within the temperature range from –40°C to +50°C.

It is worth mentioning the scales of the presented project, which is the first experience of real-time interaction between research centers in Europe and Russia aimed at developing and studying electric vehicles and their systems.

4. References

[1] Albers A, You Y, Klingler S, Behrendt M, Zhang T, Song K. 2013 *A New Validation Concept for Globally Distributed Multidisciplinary Product Development*. Proceedings of 20th International Conference on Industrial Engineering and Engineering Management, pp. 231-242.

[2] Tibba G, Malz C, Stoermer C, Nagarajan N, Zhang L, Chakraborty S. 2016 *Testing Automotive Embedded Systems under X-in-the-loop Setups*. IEEE/ACM International Conference on Computer-Aided Design (ICCAD), Austin, TX.

[3] Gao H, Zhang T, Chen H, Zhao Z, Song K. 2018 Application of the X-in-the-Loop Testing Method in the FCV Hybrid Degree Test. *Energies*, **11 (2), 433**.

[4] Moten S, Celiberti F, Grottoli M, van der Heide A, Lemmens Y. 2018 *X-in-the-loop advanced driving simulation platform for the design, development, testing and validation of ADAS*. IEEE Intelligent Vehicles Symposium (IV), Changshu, pp. 1874-1878.

[5] Schreiber V, Ivanov V, Augsburg K, Noack M, Shyrokau B, Sandu C, Els P S. 2018 Shared and Distributed X-in-the-Loop Tests for Automotive Systems: Feasibility Study. *IEEE Access*, **6**, pp. 4017-4026.

[6] Ivanov V, Savitski D. 2015 Systematization of Integrated Motion Control of Ground Vehicles. *IEEE Access*, **3**, pp. 2080-2099.

[7] Goggia T, Sorniotti A, De Novellis L, Ferrara A, Pennycott A, Gruber P, Yunus I. 2015 Integral Sliding Mode for the Yaw moment Control of Four-Wheel-Drive Fully Electric Vehicles with In-Wheel Motors. *Int. J. of Powertrains*, **4 (4)**, pp. 388-419.

[8] Halbach S, Sharer P, Pagerit S, Rousseau A, et al. 2010 Model Architecture, Methods, and Interfaces for Efficient Math-Based Design and Simulation of Automotive Control Systems. *SAE Technical Papers*, **2010-01-0241**.

[9] Michaels L, Pagerit S, Rousseau A, Sharer P, et al. 2010 Model-Based Systems Engineering and Control System Development via Virtual Hardware-in-the-Loop Simulation. *SAE Technical Papers*, **2010-01-2325**.

[10] Gopal R V, Rousseau A. 2011 System Analysis Using Multiple Expert Tools. *SAE Technical Papers*, **2011-01-0754**.

[11] Ivanov V, Augsburg K, Bernad C, Dhaens M, Dutré, M, Gramstat S, Magnin P, Schreiber V, Skrt U, Van Kelecom N. 2019 Connected and Shared X-in-the-Loop Technologies for Electric Vehicle Design. *World Electr. Veh. J.*, **10, 83**.

[12] Wang Y, Gao Q, Zhang T, Wang G, Jiang Z, Li Y. 2017 Advances in Integrated Vehicle Thermal Management and Numerical Simulation. *Energies*, **10, 1636**.

[13] Terenchenko A, Karpukhin K, Kurmaev R 2015 *Features of operation of electromobile transport in the conditions of Russia* J. Paper of EVS 28 International Electric Vehicle Symposium and Exibition, KINTEX, Korea.

[14] Kurmaev R, Umnitsyn A, Terenchenko A, Karpukhin K. 2018 Development of High-Voltage Battery with Thermostatic System for Electric Vehicles. *International Journal of Mechanical Engineering and Technology*, Vol. 9, Issue 7, pp. 1340–1346.

[15] Leighton D. 2015 Combined Fluid Loop Thermal Management for Electric Drive Vehicle Range Improvement, *SAE Int. J. Passeng. Cars - Mech. Syst*, **8(2)**, pp. 711-720.

[16] Rugh J.P. 2013 *Integrated Vehicle Thermal Management – Combining Fluid Loops in Electric Drive Vehicles.* Technical presentation NREL/PR-5400-58161, Vehicle Technologies Program Annual Merit Review.

[17] Gross-Weege C., Lichius T., Baltzer S., Abel D. 2015 *Control Design for a Thermal Hardware-in-the-Loop Test Bench for Automobile Thermal Management Systems*. IFAC-PapersOnLine, Volume 48, Issue 15, 2015, pp. 441-447.

[18] Baltzer S., Lichius T., Gissing J., Jeck P., Eckstein L., Küfen J. 2014 *Hardware-in-the-Loop (HIL) Simulation with Modelica - A Design Tool for Thermal Management Systems.* Proceedings of the 10th International Modelica Conference.

[19] Eisele M.R., Hwang Y., Radermacher R. 2011 *Small-scale dynamic test facility for automotive thermal management systems.* Vehicle Thermal Management Systems Conference and Exhibition (VTMS10), pp.523-535.

Acknowledgments
The paper was prepared under the agreement # № 075-11-2019-080 (05.628.21.0015) with the Ministry of science and higher education Russian Federation (unique project identifier RFMEFI62819X0015).

IOP Conf. Series: Earth and Environmental Science **633** (2021) 012009 doi:10.1088/1755-1315/633/1/012009

Numerical investigation of conical vortex tube with combined axial-tangential injection

M.Guen[1], A.Miloud [1], O.Imine[2]

1 Aero-Hydrodynamics laboratory P.O.B 1505 El M'naouer Oran Algeria
2 Aerodynamics and propulsion systems laboratory P.O.B 1505 El M'naouer Oran Algeria
E-mail: gustmbo.dz@gmail.com, abnaim_5@yahoo.fr

Abstract. The present work is a numerical investigation of the combined axial- tangential injection effect on conical Ranque-Hilsch vortex tube performances. An orifice in the center of the conical control valve is created, allowing the injection of compressed air axially. The computational fluid dynamics CFD commercial software ANSYS CFX has been used for the present investigation, and the mean field is determined using the Reynolds-averaged Navier-Stokes equations RANS method. The main results show that the conical configuration causes a considerable drop in temperature of the cold and hot outlet flows up to 3°C on average and a significant increase in the cold mass flow of 25% compared to the cylindrical vortex tube with axial injection.

Abbreviations

H : Total enthalpy [J/Kg]
F_1 : First SST blending function
F_2 : Second SST blending function
K : Turbulence kinetic energy
P_k : Shear production of turbulence
P : Pressure [Pa]
R : Universal gas contant [J/mol K]
t : Time [s]
T : Temperature [K]
u,v ; Velocity [m/s]
ρ : Density [Kg/m^3]
τ : Shear stress [N/m^2]
μ : Dynamic viscosity
μ_t : Turbulent viscosity
γ_t : Kinematic turbulent viscosity
ω : Turbulent frequency

1. Introduction

The Ranque–Hilsch vortex tube is a simple cylindrical device capable of dividing the incoming compressed air into two fluxes with different temperatures, cold and hot. This process is known as energy separation or the Ranque effect. Several numerical and experimental studies have been realized in this research field, with the aim of finding a scientific explanation for the temperature separation phenomenon. In 1933, the first study was proposed by Ranque [1], affirming that separation is the result of adiabatic compression of peripheral flow. In 1947, Hilsch [2] noticed that the separation phenomenon is due to the spiral motion of the incoming compressed air, which has been divided into two streams. One heated by compression near the wall, and the other cooled by expansion in the middle part of the tube. Many researchers have proposed different theories and hypotheses related to the temperature separation effect. Xue et al [3] presented a study to provide a new explanation for the energy separation inside a vortex tube. Then, exergy analysis is also carried out using current and previous experimental investigations. Li et al [4] have experimentally studied the effect of different working fluids R 134, R744, R32, R227 relative to different inlet pressures (500 Kpa- 850 Kpa), and different inlet temperature (308.15 K- 33.1 K) for a range of cold mass fractions (20% - 97%) on the vortex tube performance and temperature separation. They concluded that the working fluid R134 has great effect on the cold temperature difference and the enthalpy difference comparing with other gases. Furthermore, with the inlet pressure increase, the cold temperature also improves and the hot temperature decreases gradually. Thakare et al [5] numerically studied the influence of the turbulence models, namely one equation Spalart-Allmaras, two equations, standart K-ε and Standart K-ω model on the thermo-physical characteristic relative to different working fluids. Their main results showed that the predicted energy separation using Spalart-Allmaras model on the cold side gives better results. However, the secondary flow occurrence was detected by all turbulence models. Rafiee [6] conducted an experimental investigation on the influential parameters such as the control valve cone length, nozzle number and inlet pressure on the performance of vortex tube. The results show the existence of optimum cone length to reach minimum cold temperature, and the number of inlet nozzles is found to be the main parameter influencing swirling flow inside the tube. Farzaneh-Gord et al [7] experimentally investigated the influence of vortex generator characteristics namely; cold orifice angle, cold orifice diameter and nozzle area using a natural gas as working fluid under inlet pressure of 4 bars. The main results indicated that the cold orifice angle of 4.1°, cold orifice ratio of 0.64 and nozzle area ratio of 0.14 give the best vortex tube efficiency. Rahbar et al [8] presented a numerical investigation on the flow pattern of micro scale vortex tube. Two models were performed for 2D axisymmetric and 3D using a commercial software Fluent 6.3.26. The main results were in good agreement with experimental data of the previous study. They concluded that the 3D simulation is more efficient than the 2D numerical investigation. The maximum cold temperature drop and maximum efficiency were achieved at cold mass fraction 0.58. The maximum cold energy occurred at a cold mass fraction 0, 65. Due to the friction of fluid layers, the tangential velocity decreases along the tube after achieving the maximum value near the inlet entrance. A. Vladimir et al [9] performed a numerical study of 3D model of a double-circuit vortex tube by using commercial software ANSYS- CFX 14.5. Six turbulence models have been tested such as K-ε, - RNG - K-ε, SAS-SST, RSM-LRR and LES. The most results showed that the LES turbulence model described the flow behaviour inside vortex tube more accurately. In addition, the LES model indicates a very high radial velocity, which is the most influential parameter on the turbulent intensity and mass transfer. H.R. Thakare et al [10] carried out a numerical investigation of a 2D axisymmetric model of a counter flow vortex tube. The K-ε turbulence model was used to predict the flow behaviour inside the tube. The result indicates the existence of recirculation secondary flow near the entrance. Furthermore, the velocity vectors showed the appearance of backflow near the cold exit, which diminishes with the increase of cold mass flow fraction. G. M. P. Yadav et al [11] experimentally investigated the performance of forced-vortex flows positioned at the end of the cold exit. This new design of vortex tube is a Dual forced flow vortex tube (DFFVT). The experiment was conducted under different conditions such as inlet pressure, variable cold mass fraction and cold end diameter I and II, which have been shown to have a great effect on flow pattern and temperature drop. The maximum temperature difference is achieved with larger diameters at high cold mass fractions. In low fractions, the smaller cold diameter II attains a high temperature drop. The coefficient of performance COP achieve its maximum value of 0.057 with an orifice diameter of 4mm.V.Nikitin et al [12] presented an overview of different numerical works for simulating the energy separation and flow pattern inside vortex tube, relative to 2D axisymmetric and 3D numerical models using a different turbulence models. This analysis is focused on recent works which are in good agreement with experimental results. S. E. Rafiee et al [13] numerically studied the influence of rounding-off edge radius inside the entrance of vortex chamber in 3D vortex tube mode with the increasing of total pressure, and tangential velocity. These conditions have a major effect on the energy separation, and improve the cold temperature drop. The highest cold temperature drop is achieved at edge radius of 1,5 mm. S. N. Jadhav et al [14] presented an experimental study on the effect of working and geometrical parameter of vortex tube constructed with UPVC (Unplasticised Polyvinyl Chloride) material, which has a very low thermal conductivity. The maximum cold and hot temperature difference were reached at L/D = 50. The hot temperature increases with a decrease in the cone angle of the control valve. Furthermore, the cold temperature drop is reached with increasing inlet pressure. K. D. Divade et al [15] conducted an experimental investigation on the effect of geometrical parameters of vortex tube such as L/D ratio, exit control valve angles, number of nozzle and tube divergence angle, and using air as working fluid composed of only nitrogen, oxygen and water vapour. Operational parameters have also been investigated like inlet pressure, cold mass fraction, and Mach number (subsonic, sonic, and supersonic). The obtained results confirmed that all these parameters have considerable influence on tube performance and temperature separation. Particularly in low cold mass fraction values and subsonic Mach number, the vortex tube achieves its optimum performance. K. K. KumarRao et al [16] experimentally studied the influence of geometrical parameters on the performance of vortex tubes, using a wood tube build with two kinds of wood, rosewood and sapota wood. The main results showed that the rose wood provides slightly better performance compared to sapota wood. H. Pouraria et al [17] presented work based on Artificial Neural Networks (A.N.N) model to predict the effect of geometrical and operational parameters such as ratio Length to diameter of the tube, ratio of cold exit diameter to tube diameter inlet pressure cold mass fraction, of a vortex tube on its cooling performance. The generic algorithms help to optimize the ANN model. The results of their study show that ANN model is considered to be an adequate tool for predicting the thermodynamics parameter of the vortex tube with good precision. The vortex tube achieves optimal performance at a ratio of cold exit diameter to tube diameter d/D=0, 5.Additionally, the results show the increase in length to the diameter ratio led to an increase in the cooling performance of vortex tube. K. D. Divade et al [18] conducted an experimental investigation on vortex tube geometrical and operational parameters effect such as L/D ratio, number of nozzles, nozzle geometry, valve angles and cold orifice diameters under different inlet pressures, their results showed that the L/D = 17 provide a high COP and cold temperature drop relative to cold mass fraction equal to 0.78. A. Haydari et al [19] presented a new method to investigate the effect of fin installation 1, 3 and 5 on the hot side in order to improve the cooling performance of the vortex tube. Their results denote that the 5 triangular fins and 3 square fins provide the highest COP, isentropic efficiency and lowest cold temperature. Y. Xue [20] presented work based on flow visualization using PIV measurement on transparent vortex tube in order to understanding of the internal flow mechanism to explain the separating processes inside the vortex tube and improve the tube performance their results showed a good agreement between the observed flow and previous published studies. A. Bazgir et al [21] carried out a study in order to investigate the impact of different turbulence models on the energy separation. Ten turbulence models have been tested in steady and transient state. The result reveals that the RNG- K-ε turbulence model provides a good agreement with a experimental results. The outcomes prove that the separation energy result from the pressure loss inside the vortex tube. A. Bazgir et al [22] conducted a numerical investigation on the effect of fins installation on the cold tube including triangle, square, rectangle, circle, parallelogram and trapezium. The outcomes indicate that the parallelogram and rectangular fins have the lower cold temperature additionally the maximum values of COP and isentropic efficiency were reached. The present study focuses on the performance of a new configuration of conical vortex tube with a cone angle of 10° and axial injection characterized by the installation of a second inlet arranged axially on the hot outlet side in the conical control valve Figure.1. This orifice is geometrically identical to the tangential orifice, it is expected from this new conical vortex tube configuration that the axial injection of compressed air will accelerate the central flow inside the vortex tube, which causes a considerable increase in the cold mass flow and significantly a decreasing the temperature of both hot and cold flows. The present study was performed numerically, using the commercial software ANSYS-CFX. The dimensions of conical vortex tube are identical to the commercial vortex tube used in the experimental investigation of Ref

[23]. The main results showed that a gain of 25% has been reached in the flow of the cold flow in addition considerable drop of temperature of the two sides warm and cold estimated at 3°C compared with Ref [24].

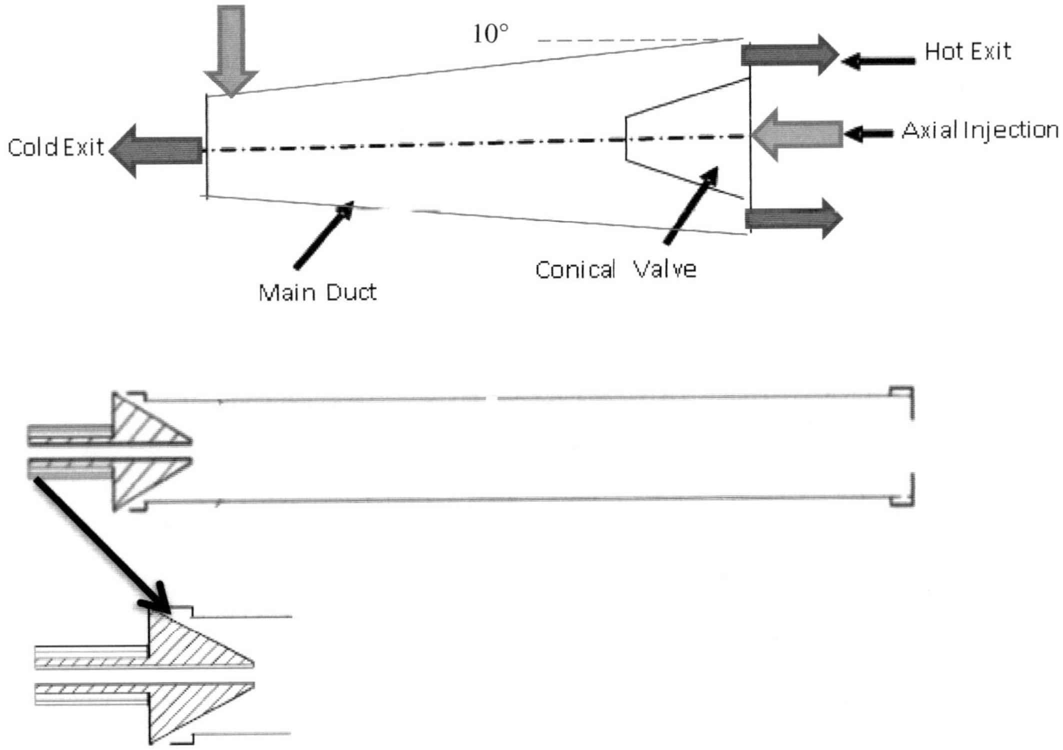

Figure 1. Schematic of Conical Vortex Tube with Axial Injection

2. Governing equation

In order to compute the flow field inside the vortex tube, some assumptions are considered:
- Compressible fluid,
- Steady and turbulent flow
- Physical proprieties expressed as temperature functions.

The governing equations including conservation of mass, momentum, energy and state equation were discretized using high-resolution schemes based on finite volume approach. These equations are:

Continuity equation

$$\frac{\partial \rho}{\partial t} + \nabla(\rho.u) = 0 \quad (1)$$

Momentum equation

$$\frac{\partial(\rho.u)}{\partial t} + \nabla(\rho u.u) = -\nabla p + \nabla(\tau) \quad (2)$$

IOP Conf. Series: Earth and Environmental Science **633** (2021) 012009 doi:10.1088/1755-1315/633/1/012009

Energy conservation equation

$$\nabla(\rho.v.H) = \nabla\left(\frac{k}{c_p}.\nabla.H\right) - \nabla(\tau.u) \quad (3)$$

State equation of ideal gas

$$p = \rho.R.T \quad (4)$$

3. Computational domain

The 3D numerical domain shown in Fig 2.conserves the same dimensions reported in Ref [25], and the only modification was made to the conical valve by creating a hole in the middle. The rotational periodicity is also applied in the present study.

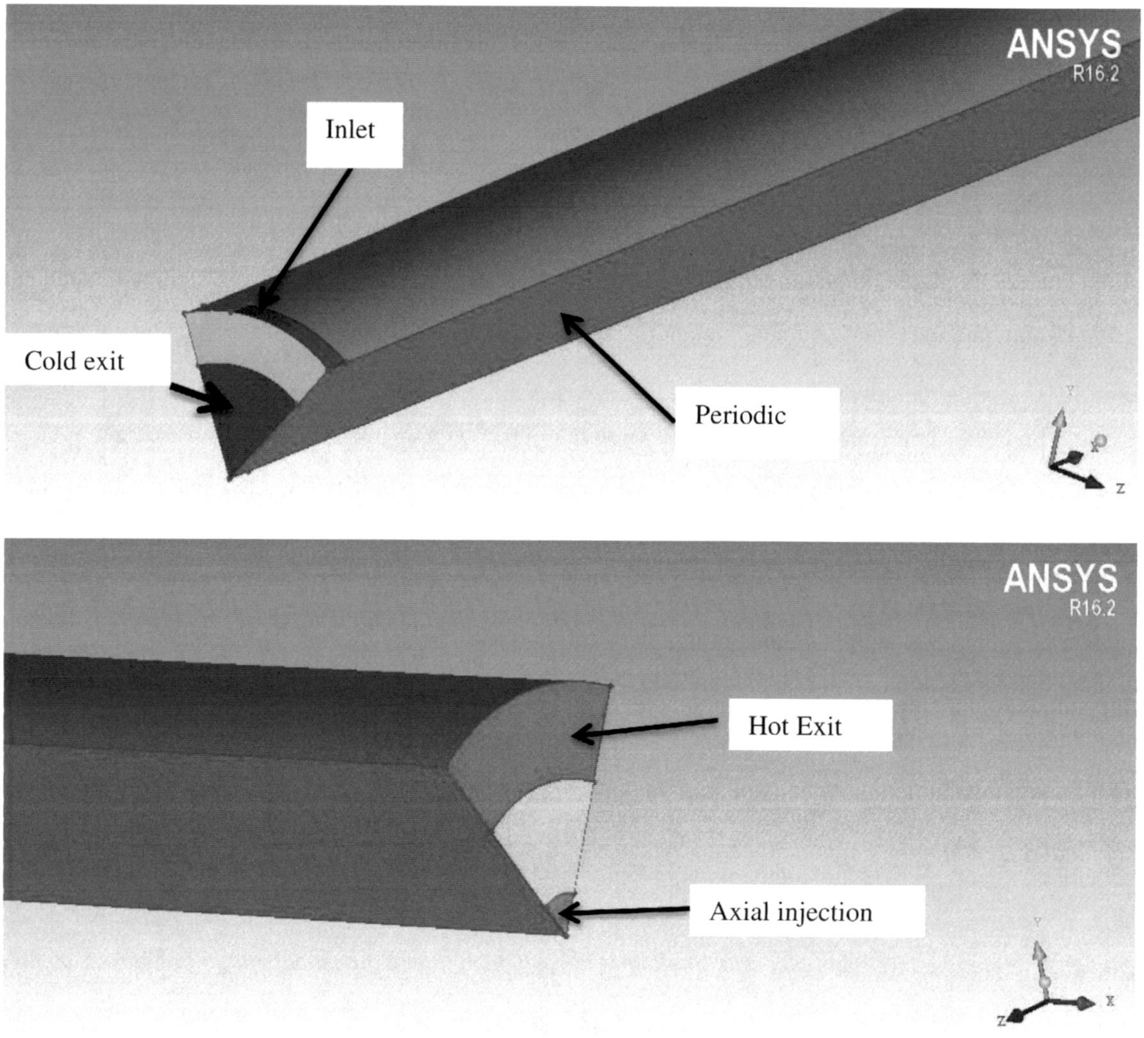

Figure 2. Numerical domain

4. Turbulence model

The flow behaviour inside vortex tube is assumed to be highly turbulent, therefore only the mean field is determined by using RANS method. Due to its ability to resolve several numerical problems related to the convergence, the K-ω SST turbulence model is applied in the present study. The transport equations are given as [25].

Turbulence kinetic energy

$$\frac{\partial}{\partial t}(\rho.k)+\frac{\partial}{\partial x_j}(\rho.U_j.k)-\frac{\partial}{\partial x_j}\left(\left(\mu+\frac{\mu_t}{\sigma_k}\right).\frac{\partial k}{\partial x_j}\right)=\tau_{ij}.\frac{\partial U_i}{\partial x_j}-\beta^*.\rho.k.\omega$$

Specific dissipation rate

$$\frac{\partial}{\partial t}(\rho.\omega)+\frac{\partial}{\partial x_j}(\rho.U_j.k)-\frac{\partial}{\partial x_j}\left(\left(\mu+\frac{\mu_t}{\sigma_\omega}\right).\frac{\partial \omega}{\partial x_j}\right)=\frac{\gamma}{\nu_t}.\tau_{ij}.\frac{\partial U_t}{\partial x_j}-\beta.\rho.\omega^2+$$

$$2.\rho.(1-F_1).\frac{1}{\sigma_{\omega_2}.\omega}.\frac{\partial k}{\partial x_j}.\frac{\partial \omega}{\partial x_j}$$

The constant values are given as

$$\sigma=F_1.\sigma_1+(1-F_1).\sigma_2$$

$$F_1=\tanh\left(\arg_1^2\right)$$

$$F_2=\tanh\left(\arg_2^2\right)$$

With

$$\arg_1=\min\left[\max\left(\frac{\sqrt{k}}{0.09.\omega_y};\frac{500.\mu}{y^2.\omega}\right);\frac{4.\rho.k}{CD_{k\omega}.y^2.\sigma_{\omega_2}}\right]$$

$$\arg_2=\min\left[\max\left(2.\frac{\sqrt{k}}{0.09.\omega_y};\frac{500.\mu}{y^2.\omega}\right);\frac{4.\rho.k}{CD_{k\omega}.y^2.\sigma_{\omega_2}}\right]$$

$$CD_{k\omega}=\max\left(2.\rho.\frac{1}{\sigma_{\omega_2}.\omega}.\frac{\partial k}{\partial x_j};10^{-20}\right)$$

The constant values are given as

$$\beta_1=0.0750,\ \gamma_1=0.5530\ \sigma_{k_1}=1.176\ \sigma_{\omega_1}=2.000\ \beta_2=0.0828\ \beta^*=0.090\ \gamma_2=0.4404$$

$$\sigma_{k_2}=1.000\ \sigma_{\omega_2}=1.170\ \alpha_1=0.31$$

5. The performance parameters

The cold mass fraction parameter is the most studied in several vortex tube studies. The majority of performance curves are drawn function of this parameter. The cold mass fraction can be expressed as

$$\varepsilon=\frac{\dot{m}_c}{\dot{m}_i}$$

$1-\varepsilon$ Hot mass fraction

Hot and cold temperature differences are given by the relationships

$$\Delta T_c=T_i-T_c$$

$$\Delta T_h=T_h-T_i$$

Using some assumptions such as neglecting the kinetic energy of the gas and considering no heat exchange to the vortex tube from its surrounding the isentropic efficiency can be expressed as [25]

IOP Conf. Series: Earth and Environmental Science **633** (2021) 012009 doi:10.1088/1755-1315/633/1/012009

$$\eta_{is} = \frac{T_{in} - T_c}{T_{in} - T_s}$$

Knowing that

$$T_s = T_{in}\left(\frac{P_c}{P_{in}}\right)^{\frac{\gamma-1}{\gamma}}$$

By replacing the two last equations the isentropic efficiency expression become

$$\eta_{is} = \frac{T_{in} - T_c}{T_{in}\left[1-\left(P_{atm}/P_{in}\right)^{\frac{\gamma-1}{\gamma}} - 1\right]}$$

6. Mesh sensitivity

The studied geometry meshing is carried out by using ICEM software. To assure greater precision, the domain is meshed by adopting hexahedral elements which insufficiently refined near the wall to represent correctly the flow behaviour in the boundary layer. In the present study, five grid cases are tested namely N= **57240** cells**, N=114480** cells, **N=130251**cells, **N=190254** cells, Fig. 3. As it is shown in the Table.1, the total temperature difference between cold and hot exits relative to different cases shows that the solution remains unchanged after **N= 170145** cells, therefore the mesh of **N= 190254** cells is adopted for the following computations.

Table 1. Total temperature difference for different grids.

Cells	114480	228960	260000	280000	.300000
ΔT	22.1	18.86	15.8	133,1	13.3

Figure 3. Mesh geometry

7. Boundary conditions

In the present work, the cold and hot exits sections are considered constant at atmospheric pressure. Therefore to modify the cold mass fraction, it is necessary to increment the hot exit pressures with 0.05 bars in each step of numerical calculation and keep the cold exit pressure constant. The mass flow rate at the tangential entrance is fixed at 8.35 g/s with circumferential angle of 42°. On the other hand, the mass flow at the axial entrance is maintained at 1.75 g/s with a section of 6,15 mm^2. The total temperature at the tangential entrance is 294.2 K

8. Numerical results

All calculations in the present study have been performed by using the commercial software ANSYS-CFX. The governing equations are discretized by using high resolution schemes. The convergence is achieved when the residues reach 10^{-7} for the continuity and 10^{-6} for the remaining equations. Figure 4 show the streamlines inside the vortex tube in the central plane The exam of figure 4 reveals that the central flow undergoes a slight deceleration, followed by acceleration. It is clear that the existence of secondary flow in neighbouring of the cold side. and two others vortices near the hot exit, which contribute to the contraction of central streamline resulting in central flow acceleration. The figure 4 gives an exemple of the spiral flow inside the vortex tube where it is remarked the acceleration of the central flow which starts at one-fifth of the length (1/6) of vortex tube from the control valve. Therefore anoticeable increase in the cold flow rate is expected in comparison to the case of classical vortex

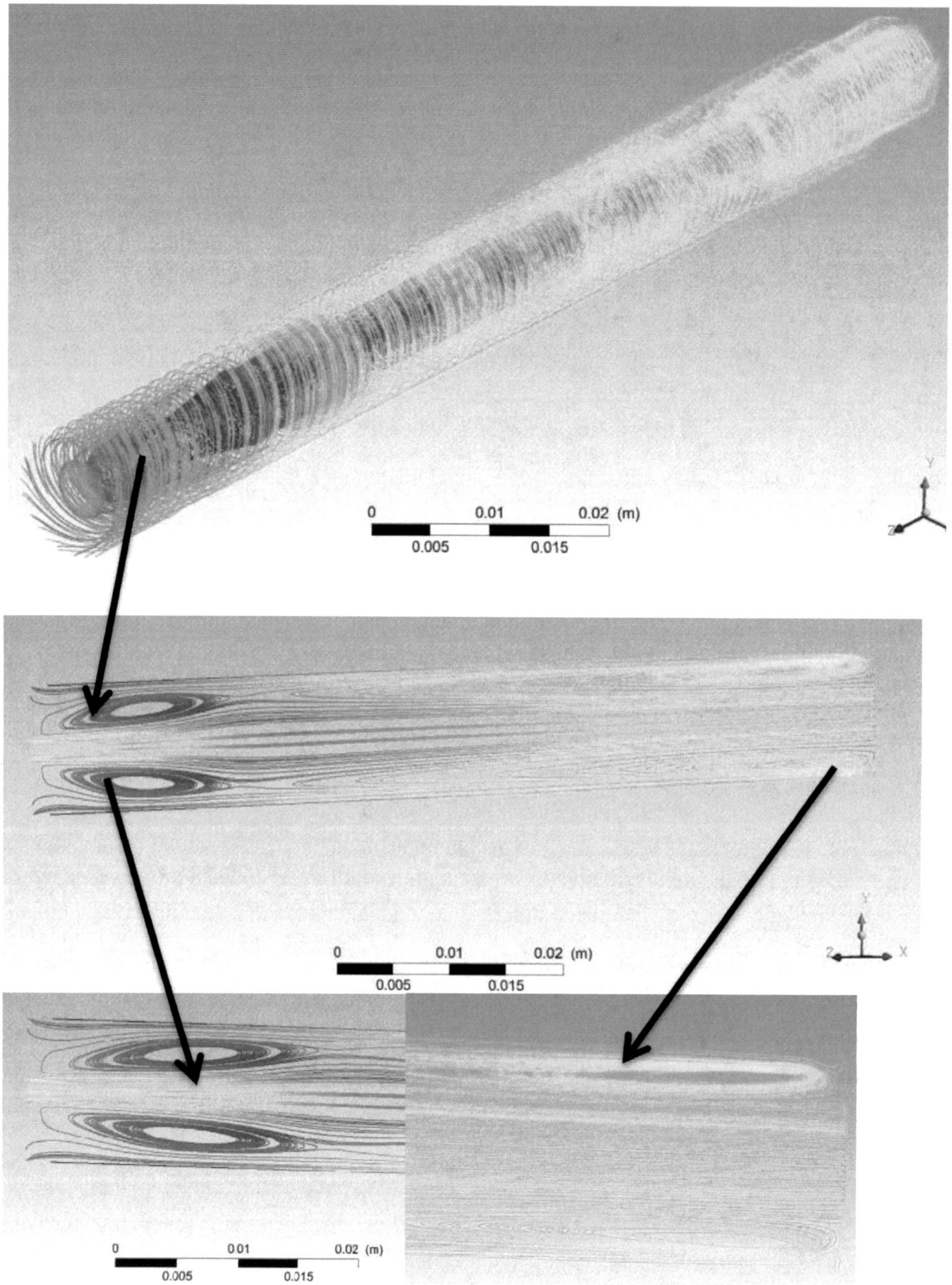
0
0.01
0.02 (m)
0.005
0.015
Y
Z
0
0.01
0.02 (m)
0.005
0.015
Y
Z
X
0
0.01
0.02 (m)
0.005
0.015

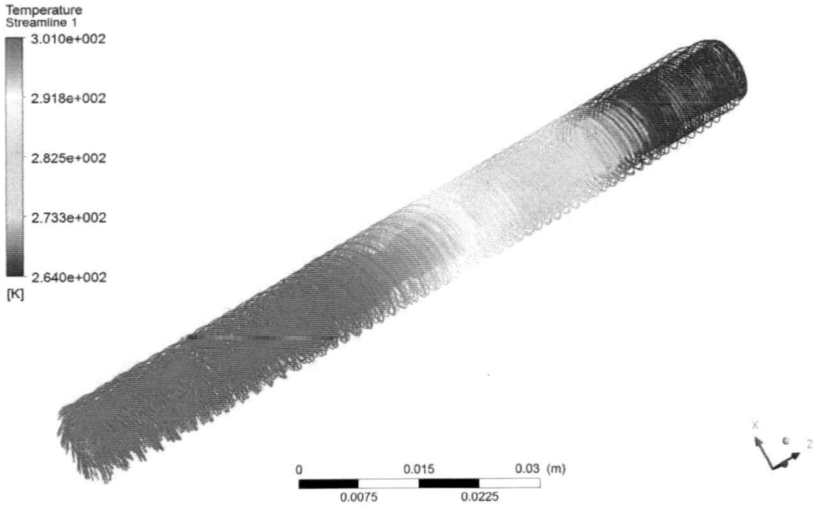

Figure 4. Flow field inside Vortex Tube

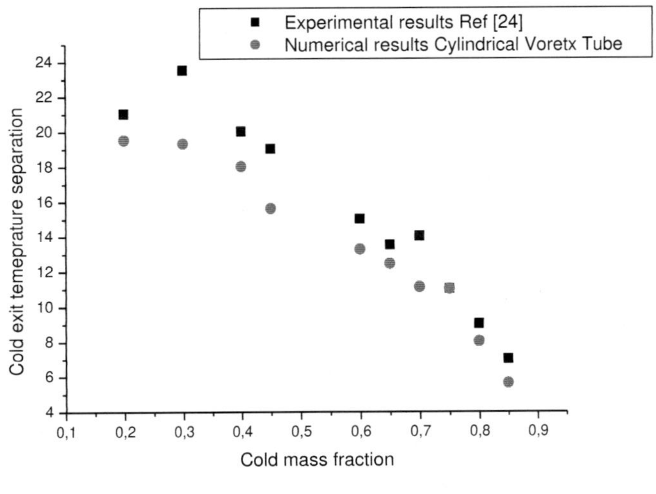

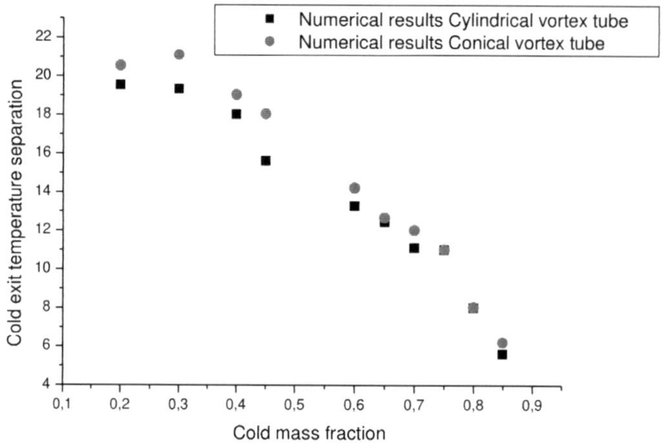

Figure 5. Cold exit Temperature Separation

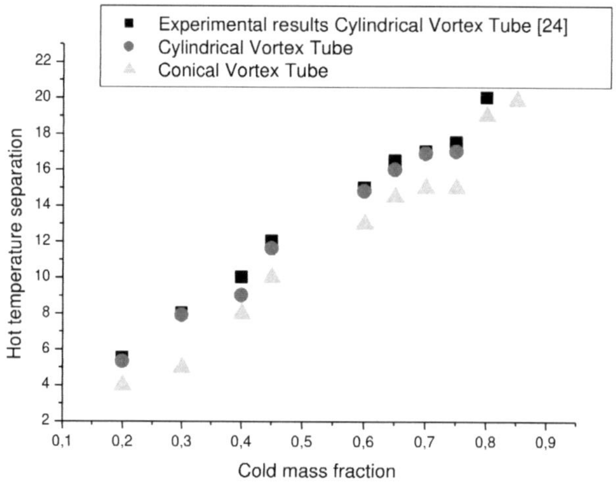

Figure 6. Hot exit Temperature Separation

Figures 5 and 6 represent the temperature separation predicted by the CFD investigation. As seen in figure 5, it is showed a significant improvement in cold temperature comparing with ordinary vortex tube. Indeed, an important drop in the cold temperature, about 3°C, is observed. However, in figure 6, it is remarked that the warm outflow through the hot exit was reduced Compared to classical vortex tubes, This effect is due to the conical shape of the tube.

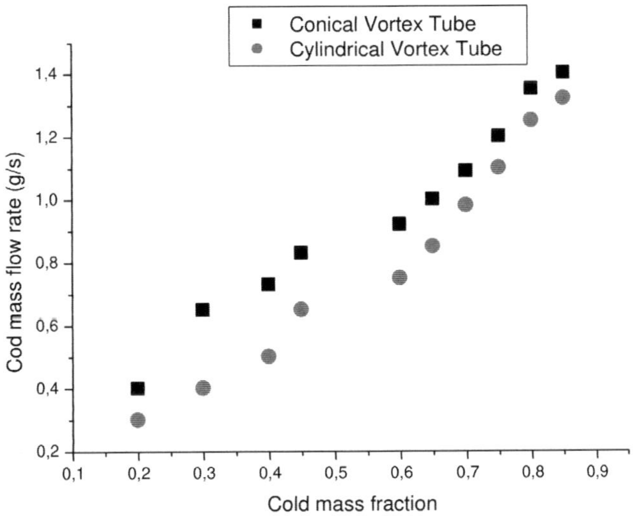

Figure 7. Cold mass flow rate

Figure 7 and figure 8 shows the cold mass flow rate predicted numerically for the two cases conical and classical vortex tube. It can be seen that there is a significant gap between the two curves. It is to note that the conical shape of the vortex tube has a significant effect on the increases of the mass flow rate of the cold side This indicates that a part of the warm flow crosses the cold exit. This is effectively confirmed by figure 8, which shows the evolution of hot mass flow rate

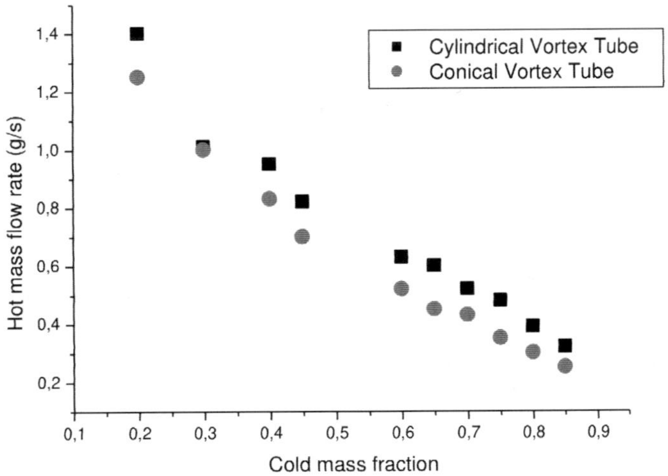

Figure 8. Hot mass flow rate

9. Conclusion

The numerical investigation of conical vortex tube with axial injection has revealed many interesting results compared to those of cylindrical vortex tube. The main findings are cited below:

- An improvement in cold temperature separation compared with ref [24]
- A significant drop in the hot temperature separation regarding to ref [24]
- An important gain of 25 % in cold mass flow rate

The present results have to be checked by carrying out an experiment series on the conical configuration of vortex tube with axial injection. Furthermore, it would be of interest to study the variation of some vortex tube thermodynamics parameters performance.

10. References

[1] GJ .Ranque (1933) Experiments bon expansion in a vortex with simultaneous exhaust of hot and cold air. Le Journal de Physique, Volume 4 1125–1130

[2] R. Hilsch (1947) The use of the expansion of gases in a centrifugal field as cooling process. Rev SciInstrum Volume 18: 108–113

[3] Y. Xue, M. Arjomandi, R. Kelso (2014) Exergy analysis within a vortex tube. Experimental Thermal and Fluid Science Vol 52

[4] N.Li, Z.Wang, X.Ham Experimental study of the coupling characteristics between vortex tube and refrigerents International Refrigeration and Air conditioning Conference at Perdue (14-17 July 2014).

[5] H.R. Thakare, A.D.Parekh (2014) CFD analysis of energy separation of vortex tube employing different gases, turbulence models and discretization Schemes. International Journal of Heat and Mass Transfer Vol 78

[6] S.E.Rafiee, M.M.Sadeghiazad (2014). Three-dimensional and experimental investigation on the effect of cone length of throttle valve on thermal performance of vortex tube using K-ε turbulence model. Applied Thermal Engineering Vol66

[7] M.Farzaneh-Gord, MeisamSadi (2014). Improving vortex tube performance based on vortex generator design. Energy Vol 72

[8] N. Rahbar, M.Taherian, M.Shateri, M.S.Valipour (2015). Numerical investigation on flow behavior and energy separation in a micro-scale vortex tube Thermal Science Volume 9

[9] A.Vladimir, V.Bianco, A.Khiat, A.Noskov (2015). Numerical investigation of a double-circuit Ranque-Hilsch vortex tube. International Journal of Thermal Sciences Volume 89

[10] H.R.Thakare, A.Monde, B.S.Patil, A.D.Parekh (2015) Numerical investigation of flow characteristics in counter flow vortex tube. Procedia Engineering Volume 127.

[11] G.M.P.Yadav, P.M.Reddy, B.U.M.Gowd (2015) Effectof cold fraction and orifice diameter on the performance of modified vortex tube with Dual forced vortex flow. International Journal of Thermal Technologies. Volume 5 .

[12] V.Nikitin, b.Paulius, B.Marijonas (2015) Overview of numerical Methods for simulating Ranque-Hilsch Effect within vortex tubes. Mokslas-LietuvosAteitis Science Future of Lithuania .

[13] S.E.Rafiee, M.M.Sadeghiazad (2015) 3D Numerical Analysis on the effect of rounding off edge radius on thermal separation inside a vortex tube. International Journal of Heat and Technology Vol 33.

[14] S.N.Jadhav, V.D.Wagh, N.Mahesh, N.Patil, V.Balaji, P.P.Patunkar (2015) An experimental Modeling and investigations of vortex tube using UPVC Material. International Journal of Technology Enhancement Engineering Research Vol 3.

[15] K.D.Devade, A.T. Pise (2016). Effect of Mach number, valve angle and length to diameter ratio on thermal performance in flow of air through Ranque-Hilsch Vortex Tube. Heat and Mass Transfer DOI 10.1007/s00231-016-1805-6

[16] K.K.KumarRao, B.V.M.Reddy, A.Ramesh, M.T.Dennes. (2016) Experimental Analysis of vortex tube made of homogeneous wood. International journal of Engineering Research & Technology Vol 5.

[17] H.Pouraria, S.M.Kia, P.Warn-Gyu, B.Mehdizadeh (2016) Modeling the cooling performance of Vortex Tube using a Generic Algorithm-Based Artificial Neural Network. Thermal Science Vol 20.

[18] K.D.Devade, A.T. Pise (2018) Parametric Analysis of thermal performance of Ranque-Hilsch vortex tube . Journal of Thermal Engineering. Vol 4 N° 5

IOP Conf. Series: Earth and Environmental Science **633** (2021) 012009 doi:10.1088/1755-1315/633/1/012009

[19] A. Haydari, A. Bazgir, M.Shahibzadeh, N.Nabhani (2019) Optimization of an environmental friendly Ranque-Hilsch Vortex Tube energy system by means of CFD simulation in Order to produce sustainable energy. Energy & Environmental Sciences.

[20] Y.Xue, J.R. Binns, M.Arjomandi, H.Yan (2019) Experimental investigation of the flow characteristics within a vortex tube with different configurations. Int Journal of Heat and fluid flow Volume 75 195-208.

[21] A.Bazgir, N.Nabhani, A.Haydari (2019) Counter flow RHVT(C-RHVT) An experimental and CFD Numerical study. LAMBERT Academic Publishing.

[22] A.Bazgir, A.Haydari, N.Nabhani (2019) Investigation of the thermal separation in a counterflow Ranque-Hilsch Vortex Tube with regard to different geometries located inside the cold tube length International communication in Heat and Mass Transfer Volume 18

[23] Skye HM, Nellis GF, Klein SA (2006) Comparison of CFD analysis to empirical data in a Commercial vortex tube. International Journal of Refrigeration 71-80.

[24] S.A.Rafiee, M.M.Sadeghiazad (2017) Improving the energical performance of vortex tubes based on comparison parallel Ranque-Hilsch and double-cicuit vortex tubes using both experimental and CFD approaches applied thermal engineering. Volume 123 1223-1236

[25] M. Guen, C. Natkaniec,J. Kammeyer,J. R. Seume,L. Adjlout, O. Imine (2013) Effect of The conical-shape vortex tube on the performance of vortex tube Journal of HMT Volume 49.

IOP Conf. Series: Earth and Environmental Science **633** (2021) 012010 doi:10.1088/1755-1315/633/1/012010

Experimental study on regeneration performance of the heat localization method for absorption air-conditioning system

Wanshi Zhang[1], Yunlei Wu[1] and Xiuwei Li[1]

1 College of Power Engineering, Nanjing University of Science and Technology, Nanjing 210094, China
E-mail: good3000best@163.com

Abstract. Due to serious energy shortage, it is urgent to reduce energy consumption from burning of fossil fuels. Driven by low-grade heat, absorption system is a promising energy conservative air-conditioning system for buildings. However, the main obstacle is the poor performance as a result of the energy waste in conventional thermal regeneration process. With the aim of performance improvement, a heat localization method is proposed for the regeneration process of absorption system. It can improve energy utilization efficiency through thermal energy localization on the evaporation surface. Preliminary experiments have been conducted to test the regeneration performance with this new method. Some parameters have been investigated with the experimental data. Moreover, by comparison between the heat localization method and traditional heating method, it indicates the energy utilization efficiency with the heat localization method are 1.8 to 2.8 times higher than that with the traditional heating method. The performance of the system has been greatly improved. The energy utilization efficiency of the regeneration process can reach 36.9% under certain working conditions. With the performance improvement, the popularization of absorption system will be promoted.

1. Introduction

Energy is essential for humans and the energy shortage has limited the development of economy and society. With the demand increase on energy, efficient utilization of energy is not an unattractive option to solve the increasingly serious energy problem. Buildings account for about 40% of the world's total energy consumption while the air-conditioning system contributes to the largest share of building energy consumption [1], [2]. As a result, an efficient air-conditioning system is critical for energy problem-solving. Nowadays the vapor compression air-conditioning system has been widely used because of its high performance. However, it is driven by electric power and brings a great burden to the grid. Therefore, a promising alternative is absorption air-conditioning system, which can be driven by low-grade heat. The low-grade heat can be obtained from renewable energy (e.g. solar energy) and industrial waste heat (e.g. flue gas from coke ovens) [3], [4]. Nevertheless, the main drawback is that the performance of absorption system is relatively low compared with vapor compression system. It is mainly because too much energy has been wasted in conventional thermal regeneration process. After regeneration process, absorbent solution is relatively hot because it contains much unutilized thermal energy. A large portion of the input energy is used for heating absorbent solution rather than regeneration. Moreover, the high temperature of absorbent solution is harm to the following absorption process. So regenerated absorbent solution needs to be cooled down to a low temperature. The thermal energy contained in regenerated absorbent solution not only reduces the energy utilization efficiency in regeneration process but also needs to be removed by consuming excess energy, which leads to a poor performance of the whole system. To solve this problem, the membrane regeneration method and the capacitive deionization regeneration method have been used to improve the regeneration performance of absorption system [5], [6]. The regeneration process can be essentially considered as a desalination process, so these two methods can be applied to the regeneration process. Besides, the heat localization method is an appropriate method as well.

Chen et al. [7] proposed a solar steam generation approach and corresponding material structure using heat localization. Through heat localization method, thermal energy is localized on the evaporation surface and thus the heat loss to bulk liquid can be minimized [8]. In other words, the heating of bulk liquid is significantly reduced, and the temperature of bulk liquid can maintain at a low level. The decrease of energy waste will lead to a high energy utilization efficiency [9]. In recent years, many researches on heat localization method have been made: Kaur et al. [10] have developed an all-ceramic microfibrous solar steam generator to obtain clean water efficiently. Xia et al. [11] have proposed a new type of porous photothermal material based on halloysite nanotubes and graphene oxide mixture composite materials for efficient solar steam generation.

The heat localization method localizes thermal energy on the evaporation surface and thus minimizes the heat loss to bulk liquid. On account of the energy waste in conventional thermal regeneration process, the performance of absorption system is poor. With the use of heat localization method, absorbent solution can keep a low temperature after regeneration. It not only reduces the thermal energy contained in regenerated absorbent solution but also avoids cooling of regenerated absorbent solution, resulting in a higher performance of the whole system. Up till now, previous researches on heat localization method mainly concentrate on materials with high solar absorption ability and corresponding experiments are almost based on pure water [12], [13]. There is no research on absorbent solution based on heat localization method, which is critical for absorption system. The absorbent solution is different from pure water and previous research conclusions are not applicable for absorbent solution. The evaporation of water from solution is harder than from pure water and solution has stronger absorption capability with the increasing concentration. In addition, no one apply the heat localization method to other energy sources except solar energy. The change of energy source can make a difference to corresponding experimental device, which must be changed to fit the alternative energy source. They have the same principle, but the actual performance is unknown. So it is meaningful to reveal the actual performance of absorption system with heat localization method.

In this paper, we propose and test a heat localization method based absorption system. The regeneration process can be driven by low-grade heat source. For convenience, electric power is used to simulate the heat source. The experiments based on heat localization method and traditional heating method have been conducted respectively. In terms of experimental data, preliminary analysis has been made. Compared with traditional heating method, regeneration process with heat localization method has a higher energy utilization efficiency. The high regeneration performance of absorption system makes it a promising air-conditioning system.

2. Material and method

2.1. Principle of the absorption air-conditioning system

The absorption air-conditioning system is one of the oldest air-conditioning systems, which can be driven by low-grade heat. Figure 1 presents the schematic diagram of traditional absorption air-conditioning system. The main process of this system is: when adequate heat is added to the generator, the strong solution and the refrigerant vapor are obtained in regeneration process. Then the refrigerant vapor is condensed in the condenser, and the liquid refrigerant expands to a low-pressure level through the expansion valve. Next, the liquid refrigerant is evaporated in the evaporator and takes heat away to achieve refrigeration. Meanwhile, the acquired strong solution from generator is cooled in solution heat exchanger, and then expands to a low-pressure level through the expansion valve. After that, the strong solution is sent to the absorber. It is diluted to the weak solution as a result of the absorption of refrigerant vapor. Before return to generator, it is heated preliminarily in solution heat exchanger. The above cycle can repeat to achieve refrigeration continuously.

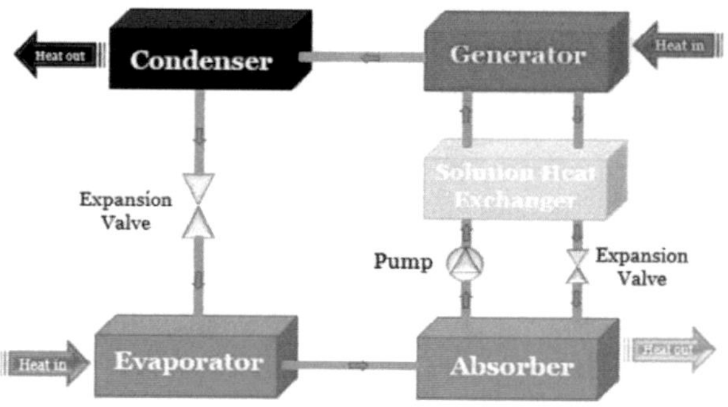

Figure 1. Schematic diagram of traditional absorption air-conditioning system [14].

The regeneration of absorbent solution costs the most energy in absorption system, which can be driven by low-grade heat source. However, too much energy has been wasted in conventional thermal regeneration process, which causes a poor performance of the whole system. The reason for that is because absorbent solution is relatively hot after regeneration, which means the input energy is not used to the full. In addition, the hot absorbent solution needs to be cooled down to a low temperature because the absorbent solution with lower temperature has a better absorption ability. The thermal energy contained in regenerated absorbent solution not only reduces the energy utilization efficiency in regeneration process but also needs to be removed by consuming excess energy. Therefore, it is necessary to improve the regeneration process of absorption system.

2.2. Principle of the heat localization method

Chen and his colleagues first proposed efficient solar steam generation approach with the use of heat localization [7]. The heat localization method brings a high energy utilization efficiency of solar steam generation because it localizes the thermal energy on the evaporation surface, which minimizes the heat loss to bulk liquid [8]. The temperature of bulk liquid can maintain at a low level during evaporation and thus the energy waste in bulk liquid is reduced. The representative heat localization structure is composed of a solar absorber layer and a thermal insulation layer (figure 2). The solar absorber layer absorbs incident solar light and converts it into thermal energy for evaporation. The thermal insulation layer should have a low thermal conductivity, which reduces the heat loss to bulk liquid. There are many channels in both the solar absorber layer and the thermal insulation layer, which wick adequate liquid from the bulk liquid to the evaporation surface under capillary action for continuous steam generation. In such a structure, the solar energy utilization efficiency can reach 84.6% at a light density of 1 kW/m^2 [15]. The heat localization method will have a great influence in many practical applications including absorption system.

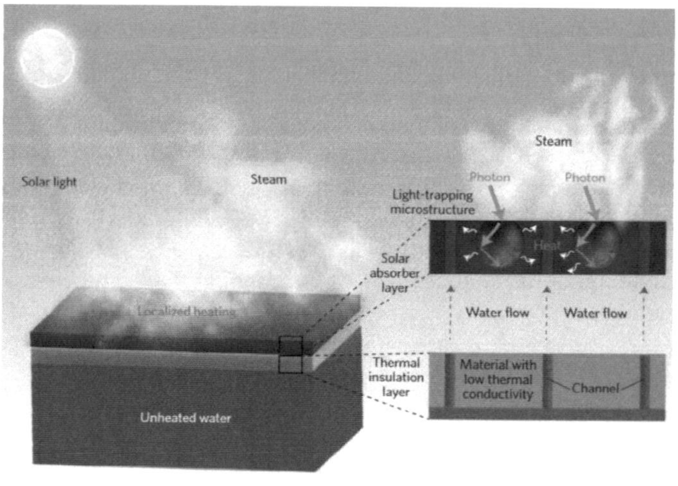

Figure 2. Schematic diagram of the representative heat localization structure [8].

The heat localization method can achieve efficient utilization of energy by localizing the thermal energy on the evaporation surface. This method reduces energy waste in heating bulk liquid and maintains bulk liquid at a low temperature during evaporation. It can be applied to the regeneration process of absorption system. Through heat localization method, the temperature of regenerated absorbent solution can maintain at a low level. The heat localization method not only reduces energy waste in heating absorbent solution but also avoids excess energy input to cool regenerated absorbent solution, which will significantly improve the performance of absorption system.

2.3. Experimental system

The regeneration process of absorbent solution costs the most energy of absorption system, which plays a vital role in the whole system. Natural wood possesses many advantages for application in heat localization structure, including excellent hydrophilicity, self-floating ability, natural microchannels and low thermal conductivity [16], [17]. We used cunninghamia as the basis of heat localization structure. The process of making heat localization structure involves two steps: wood was cut into cylindrical piece with 4 cm diameter. Then the electric heating wire as the heat source was uniformly distributed onto a cunninghamia cylindrical piece which acts as the thermal insulation layer and the liquid transport channel. The heat localization structure was showed in figure 3. In the heat localization regeneration device, heat localization structure was placed in the beaker with the volume of 20ml absorbent solution (figure 4). We chose LiCl solution as absorbent solution and tested the regeneration performance with different concentrations. The mass of absorbent solid and deionized water were measured with a high-accuracy balance (FA1004, Liangping Co., Ltd), whose accuracy was ±0.5%. It was also used to detect the solution mass change in regeneration process, which was recorded every ten seconds in fifty minutes. The energy input was 1.5W, which was supplied by a DC regulated energy source (MP1350D, Maihao Co., Ltd) with the accuracy of ±0.5%. The experiment with traditional heating method has been conducted to compare with that with heat localization method. Compared with heat localization method, the major change is that the heat localization structure is replaced by electric heating wire. It directly heated the whole absorbent solution in regeneration process. With these experimental devices, the actual evaporation rate in regeneration process has been acquired and thus energy utilization efficiency of these two methods can be obtained.

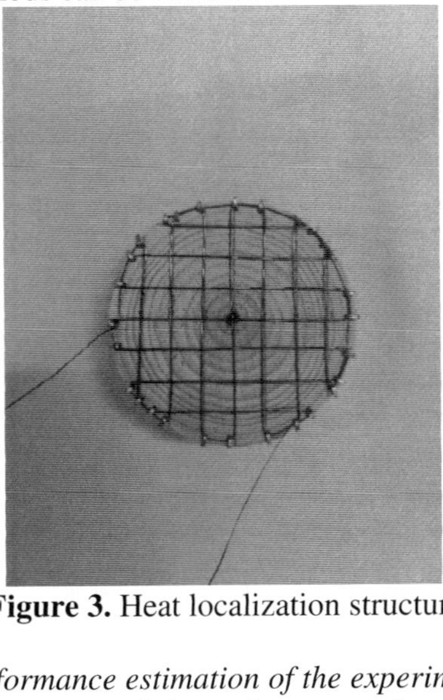

Figure 3. Heat localization structure.

Figure 4. Heat localization regeneration device.

2.4. Performance estimation of the experimental system

The performance of absorption system largely depends on regeneration process, which can be measured in terms of energy utilization efficiency. The energy utilization efficiency η can be calculated as:

$$\eta = \frac{Q}{P} = \frac{lm}{UI} \tag{1}$$

Q is the utilization of energy of the experimental system, l is the latent heat of water vaporization, and m is the evaporation rate of water. P is the input power of the experimental system, U is the voltage, and I is the current intensity. The energy utilization efficiency of the experimental system can be calculated by using equation (1).

3. Results and discussion

3.1. Experimental results

The regeneration process has great impact on the performance of absorption system. A series of experiments have been carried out to test the actual performance of regeneration process. Figure 5(a) presents the evaporation rate of LiCl solution with the mass concentration of 25%. The evaporation rate with the heat localization method did not have obvious change, which kept stable after a slight increase at the beginning. Correspondingly, the evaporation rate with traditional heating method increased significantly in the first 25 minutes, then it became stable. Overall, the evaporation rate with heat localization method was higher from beginning to end. The evaporation rate of LiCl solution (the mass concentrations are 25%, 30%, 35%) are shown in figure 5. The evaporation rate of these two methods appeared significantly drop with the increasing concentration. The reason for that is because the solution has stronger absorption capability with the increasing concentration, so it is harder to evaporate water from solution. Even in this case, it was obvious that the evaporation rate with heat localization method was higher than that with corresponding traditional heating method.

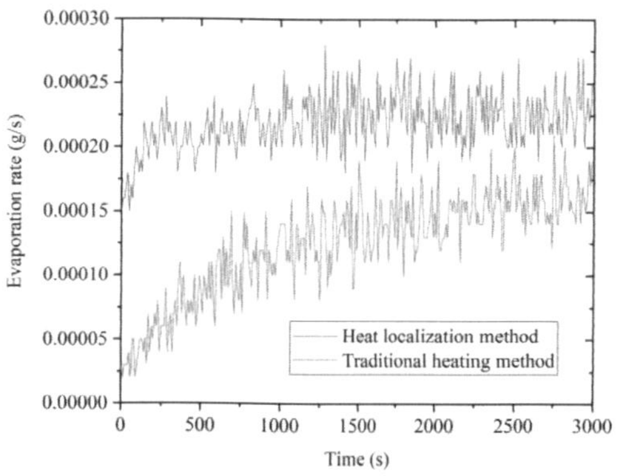

(a) mass concentration of 25%

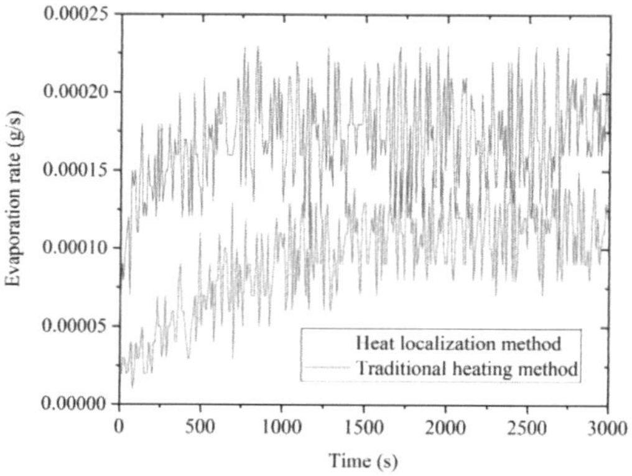

(b) mass concentration of 30%

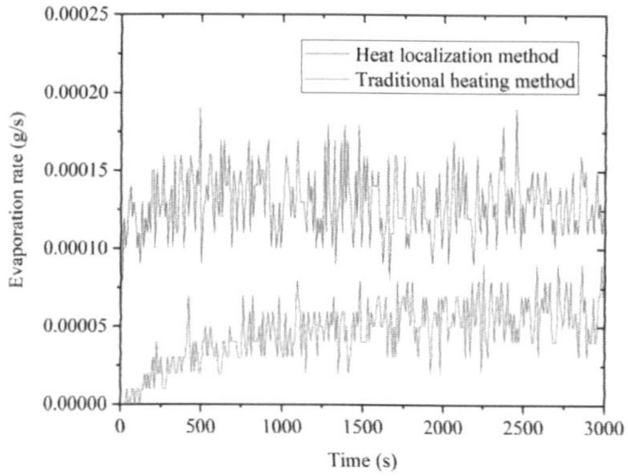

(c) mass concentration of 35%

Figure 5. The evaporation rate of LiCl solution with different concentrations.

3.2. Energy utilization efficiency of the experimental system

The energy utilization efficiency can reflect the performance of regeneration process, which has a great impact on the performance of absorption system. The energy utilization efficiency of LiCl solution is calculated and the results is shown in figure 6. The change rules and reasons of energy utilization efficiency were the same as that of evaporation rate. The energy utilization efficiency with heat localization method was higher than that with traditional heating method. The energy utilization efficiency with both methods decreased with the increasing concentration of solution. When the regeneration process was stable, the energy utilization efficiency with heat localization method reached 36.9% with the mass concentration of 25%. The average energy utilization efficiency with heat localization and traditional heating method were 35.9% and 20.4% respectively. Compared with traditional heating method, the average energy utilization efficiency improvement achieved 76.3%. When the mass concentration was 35%, the average energy utilization efficiency with heat localization method was only 21.2%. However, the improvement also achieved 180.1%.

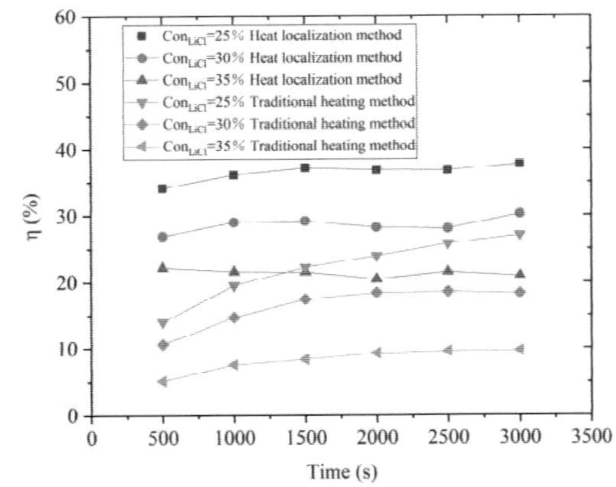

Figure 6. Energy utilization efficiency of the experimental system.

3.3. Discussion

The performance of traditional absorption system is poor because of energy waste in conventional thermal regeneration process. This problem can be solved by applying the heat localization method in regeneration process. The heat localization method in regeneration process has a high energy utilization efficiency. The reason for that is mainly because the thermal energy is localized on the evaporation surface and thus reduces the heat loss to bulk liquid. The temperature of regenerated absorbent solution can maintain at a low level, which not only reduces energy waste in heating absorbent solution but also avoids the cooling of regenerated absorbent solution.

4. Conclusions

Experimental researches of the regeneration process have been made with heat localization and traditional heating method. The results indicate heat localization method greatly improves the energy utilization efficiency, which means the performance of the whole system is further improved. When the regeneration process was stable, the energy utilization efficiency with heat localization method can reach 36.9% with the mass concentration of 25%. The energy utilization efficiency with heat localization method are 1.8 to 2.8 times higher than that with traditional heating method.

The regeneration performance of absorption system is significantly improved with heat localization method. The heat localization method has been proved to be an effective method for absorption system. Further works will focus on finding the most suitable absorbent for heat localization method.

5. References

[1] Nejat P, Jomehzadeh F, Taheri M M, Gohari M and Majid M Z A 2015 A global review of energy consumption, CO_2 emissions and policy in the residential sector (with an overview of the top ten CO_2 emitting countries) *Renewable and Sustainable Energy Reviews* **43** 843-862

[2] Chen Y, Liu Y, Liu J, Luo X, Wang D, Wang Y and Liu J 2020 Design and adaptability of photovoltaic air conditioning system based on office buildings *Solar Energy* **202** 17-24

[3] Ibrahim N I, Al-Sulaiman F A, Saat A, Rehman S and Ani F N 2020 Charging and discharging characteristics of absorption energy storage integrated with a solar driven double-effect absorption chiller for air conditioning applications *Journal of Energy Storage* **29**
[4] Sun F, Li J, Fu L, Li Y, Wang R and Zhang S 2020 New configurations of district heating and cooling system based on absorption and compression chillers driven by waste heat of flue gas from coke ovens *Energy* **193**
[5] Ding B Q, Cheng F, Zhang X S and Li X W 2018 Research on membrane absorption air-conditioning system with more than one stage *Applied Thermal Engineering* **140** 341-350
[6] Zhao H, Zhang X S, Li X W, Ding B Q and Cheng F 2018 Research on actual performance and energy recovery characteristic of capacitive deionization regeneration method for absorption air-conditioning system *Energy Conversion and Management* **171** 1549-1559
[7] Ghasemi H, Ni G, Marconnet A M, Loomis J, Yerci S, Miljkovic N and Chen G 2014 Solar steam generation by heat localization *Nature Communications*
[8] Shang W and Deng T 2016 Solar steam generation: steam by thermal concentration *Nature Energy*
[9] Liu S, Huang C, Luo X and Rao Z 2018 High-performance solar steam generation of a paper-based carbon particle system *Applied Thermal Engineering* **142** 566-572
[10] Kaur M, Ishii S, Shinde S L and Nagao T 2017 All-ceramic microfibrous solar steam generator: TiN plasmonic nanoparticle-loaded transparent microfibers *ACS Sustainable Chemistry & Engineering* **5** 8523-8528
[11] Xia M, Chen L, Zhang C, Hasi Q M, Li Z and Li H 2020 Porous architectures based on halloysite nanotubes as photothermal materials for efficient solar steam generation *Applied Clay Science* **189**
[12] Kim C, Ryu Y, Shin D, Urbas A M and Kim K 2020 Efficient solar steam generation by using metal-versatile hierarchical nanostructures for nickel and gold with aerogel insulator *Applied Surface Science* **517**
[13] Cheng G, Wang X, Liu X, He Y and Balakin B V 2019 Enhanced interfacial solar steam generation with composite reduced graphene oxide membrane *Solar Energy* **194** 415-430
[14] Nikbakhti R, Wang X, Hussein A K and Iranmanesh A 2020 Absorption cooling systems – review of various techniques for energy performance enhancement *Alexandria Engineering Journal* **59** 707-738
[15] Miao E D, Ye M Q, Guo C L, Liang L, Liu Q and Rao Z H 2019 Enhanced solar steam generation using carbon nanotube membrane distillation device with heat localization *Applied Thermal Engineering* **149** 1255-1264
[16] Xue G, Liu K, Chen Q, Yang P, Li J, Ding T, Duan J, Qi B and Zhou J 2017 Robust and low-cost flame-treated wood for high-performance solar steam generation *ACS Applied Materials & Interfaces* **9** 15052-15057
[17] Jia C, Li Y, Yang Z, Chen G, Yao Y, Jiang F, Kuang Y, Pastel G, Xie H, Yang B, Das S and Hu L 2017 Rich mesostructures derived from natural woods for solar steam generation *Joule* **1** 588-599

Acknowledgments
This research was supported by the fund of National Natural Science Foundation of China under the contract No. 51676098. It was also supported by the grants from the fund of Natural Science Foundation of Jiangsu Province under the contracts No. BK20170095 and No. BK20160822. These supports are gratefully acknowledged.

Quantification of Citarum watershed damage for flood control efficiency

Ariani Budi Safarina[1], Chairunnisa[1], Roni Farfian[1], Waluyo Hatmoko[2], Muthiah Sadidah[3]

1 Civil Engineering Department, Universitas Jenderal Achmad Yani, Jalan Terusan Jenderal Sudirman PO BOX 148, Cimahi, 40511, Indonesia
2 Research and Development Center for Water Resources, Indonesian Ministry of Public Works and Housing, Jalan Ir.H Juanda 193 Bandung, 40135, Indonesia
3 Civil Engineering Department, Institut Teknologi Bandung, Jalan Ganesha 10, Bandung, 40116, Indonesia
E-mail: ariani.budi@lecture.unjani.ac.id

Abstract. Watersheds damage causes disturbance to the river system and floods. Watershed performance can be monitored regularly through parameters to avoid damage. The purpose of this study is to formulate a quantification of watershed damage based on the flood parameters. This research was conducted in the Citarum River Basin located in West Java Province, Indonesia. The method used is an analysis of the rainfall-runoff model on four Citarum tributaries, namely Citarik, Cisangkuy, Cilember, and Cimahi River. The unit hydrograph resulting from modeling is derived to load discharge using the convolution method. Effective rainfall is determined based on runoff coefficients derived from land-use maps. The discharge capacity of the river is determined by measuring the cross-section of the river near the branching location. Indicator of watershed damage is the ratio between load discharge (Q_{load}) and river capacity discharge (Q_{cap}). The watershed is categorized as damaged if $Q_{load} / Q_{cap} > 1$. The level of watershed damage is measured by the damage index $I_D = \Delta Q / Q_{cap}$. Damage categories are low, moderate, and severe. Based on the results of the study, Cimahi river area is not damaged. Citarik River and Cisangkuy River are categorized as low damaged and Cilember River is heavily damaged.

1. Introduction

The problem of the Citarum river has a very broad impact on various levels of society. Damage of the Citarum watershed includes the amount of critical land reaching 79,668.25 ha and causing sedimentation of up to 8,465 tons/year. The sources of Citarum waters pollution include pollution from agricultural and animal husbandry activities, domestic and industrial waste, and pollution from floating net cage (KJA) aquaculture activities that are developing rapidly in reservoir waters [1]. the problems with damage to the Citarum watershed are flooding, erosion and misuse of land use. This Citarum watershed is a vital resource used by many stakeholders. Around 16% of the population of West Java is along the Upper Citarum Watershed with a population growth rate of 2.3% [2].

In the upstream area of the Citarum River, flooding often occurs in 12 Rancaekek villages with five villages experiencing the heaviest floods namely Linggar, Bojongloa, Bojongpulus, Jelekong and Sukamanah villages which are floods with heights of more than one meter and an average inundation duration of fewer than 24 hours. Floods that occur on the road that connects Bandung-Garut still occur even though there is no overflow from the Cikijing river, a tributary of the Citarik river. The efforts being carried out by stakeholders are dredging, structuring local drainage and making communal infiltration wells.

Rancaekek District is a meeting place of the Citarik river with the Citarum river. This region is also the location where the Cikijing and Cimande rivers meet the Citarik river. River branching like this is also the cause of flooding in the Cimahi city of Melong, namely the meeting of the Cilember river with the Citarum river [3]. The middle part downstream of a river area is generally a flood-prone area if no integrated management that integrates the river area and the urban regional area. This happens because the middle part of a river area will receive the impact of increased river discharge due to changes in land cover in the upstream area without spatial planning referring to the planned regional layout plan [4].

Changes in land use provide a dominant influence factor on water management and groundwater conservation which has an affects on increasing drainage coefficient [5].

Prevention of flood hazards in the province of Izmir, Turkey, is done by detecting potential flood areas using radar with five parameters and qualification levels namely very low, low, medium, high and very high [6]. The Flood Risk Index map was developed to estimate flood hazard and vulnerability in the Bukit Duri sub-district of Tebet, Jakarta, upstream of the Manggarai floodgate. The flood hazard index was analyzed based on inundation maps that were verified with field data. Based on the research risk map, the study area is a high flood risk area due to dense population housing and inadequate flood mitigation [7].

2. Methods of the study

The method used in this study is an analysis of the rainfall-runoff model on four Citarum tributaries, namely Citarik, Cisangkuy, Cilember and Cimahi river. The unit hydrograph resulting from modeling is derived to load discharge using the convolution method. Effective rainfall is determined based on runoff coefficients derived from land-use maps. The discharge capacity of the river is determined by measuring the cross-section of the river near the branching location.

Watershed damage occurs due to the disruption of the hydrology system due to changes in land use. The elements of land use that have changed generally are forests, agricultural land and settlements. The Sumpur Singkarak watershed, West Sumatra is addressing increasing land-use change by monitoring using satellite imagery interpretation and forest protection [8]. Floods in the Palu river occur because the capacity of the river is not able to accommodate the discharge load of more than 550 m^3 / s where the discharge occurs for a return period of 25 years. If the discharge load exceeds that number, floods occur. The decline in the capacity of the Palu river is due to sedimentation in the riverbed which results in aggradation [9]. Recent floods have occurred in Jayapura, Paniai and Manokwari. The cause of increased runoff during the rainy season is the conversion of forests into oil palm plantations. The solution is a land conservation program, sediment dredging and building a retaining wall [10].

Flooding is caused by the power of water damage. In general, the definition of flood is the overflow of river water so that the water flows in residential land and causes disaster [11]. Flood disaster according to the level of damage is a function of the extent of the inundation area, the length of inundation time and inundation height. Research on flooding continues to grow, starting from river flow analysis, analysis of water runoff on the ground, analysis of runoff routing mechanisms and analysis that integrates water quantity with characteristics.

A watershed can be assumed to be a lumped linear system in which the amount of water stored in a hydrological system is associated with inflow and outflow [12]. The main physical characteristics of a watershed are area, shape, elevation, slope, orientation, soil type, drainage network, water storage capacity and land cover. The effect of these types of characteristics varies. Soil types can control infiltration, surface water reservoirs, and groundwater. The combined effect of all factors is the classification for small and large watersheds [13].

The rainfall-runoff model used in this study is Nakayasu synthetic unit hydrograph method that has been calibrated based on the characteristics of the watershed [14],

$$Q_p = \frac{0.99 A R_o}{3{,}7(0{,}1 T_p + T_{0,3})} \tag{1}$$

where Peak discharge (Qp) is the function of watershed area (A), watershed characteristic coefficient (C = 0.99), unit rainfall (R_o), time lag (T_p) and time required to discharge reduction up to 30% peak discharge ($T_{0.3}$). The direct runoff discharge in each river is determined by the convolution method which is the function of effective rainfall and unit hydrograph. Discrete convolution equation for the linear system used is shown in equation 2,

$$Qn = \sum_{m=1}^{n \le m} P_m U_{n-m+1} \tag{2}$$

where Q is direct runoff discharge, P is effective rainfall and U is unit hydrograph ordinate.

The land use of a watershed shows the type of land cover of the watershed. Land surface affects the abstraction into the soil and becomes a determining variable in the amount of drainage coefficient or runoff coefficient [15][16],

$$C = \frac{r_d}{\sum_{m=1} R_m}$$

$$\sum_{m=1}^{M} R_m = \text{total rainfall} \tag{3}$$

where C is runoff coefficient and r_d is the runoff depth

The indicator of watershed damage is the ratio between load discharge (Q_{load}) and river capacity discharge (Q_{cap}). The load discharge is the direct runoff discharge coupled with the base flow of the river. The watershed is categorized as damaged if

$$\frac{Q_{load}}{Q_{cap}} > 1 \tag{4}$$

The level of watershed damage is measured by the damage index I_D

$$I_D = \frac{Q_{load} - Q_{cap}}{Q_{cap}} = \frac{\Delta Q}{Q_{cap}} \tag{5}$$

Damage categories are low ($I_D < 0.2$), moderate ($0.2 \le I_D < 0.6$) and severe ($I_D \ge 0.6$).

3. Results and discussion

3.1. Study area

Geographically, the Citarum watershed is located at 106° 57 '51 "- 107° 56' 59" East and 5o 54 '53 "- 7° 14' 38.517" S with the area of 6596 km^2. The length of the Citarum river is 297 km with the headwaters at Cisanti Lake which is located at the foot of Mount Wayang, Bandung Regency and empties into the North Coast of Java Island, Muara Gembong, Bekasi Regency. The flow of the Citarum watershed crosses 11 cities including Bandung Regency, West Bandung Regency, Purwakarta Regency, Karawang Regency, Bekasi Regency, Bandung City, Cimahi City, part of Sumedang Regency, part of Cianjur Regency, part of Bogor Regency, and part of Garut Regency. In addition to being a source of raw water for drinking water, the Citarum River also being an irrigation water source for hundreds of thousands of hectares of rice fields and power plants for Java and Bali. In the Citarum river, there are three large reservoirs, namely Saguling, Cirata, and Jatiluhur. The map of Citarum watershed is shown in figure 1,

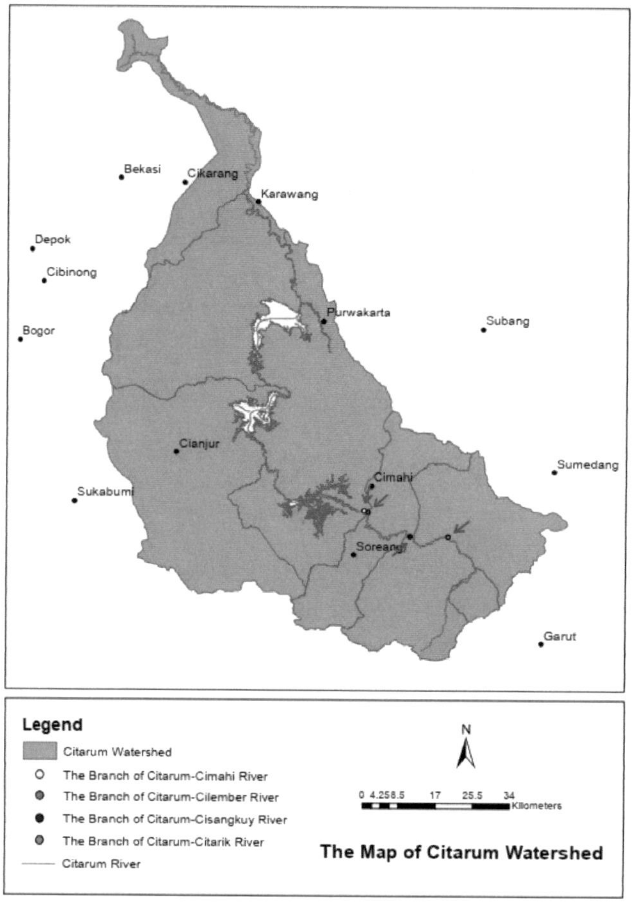

Figure 1. Citarum River Basin map

Based on the land use map, the runoff coefficient of Cimahi, Cilember, Citarik and, Cisangkuy river is 0.57, 0.57, 0.51 and, 0.61 respectively.

3.2. Observed cross-section of the river

The cross-sections of rivers in this study which the location are near the branch of Citarum are shown in Figure 2 below,

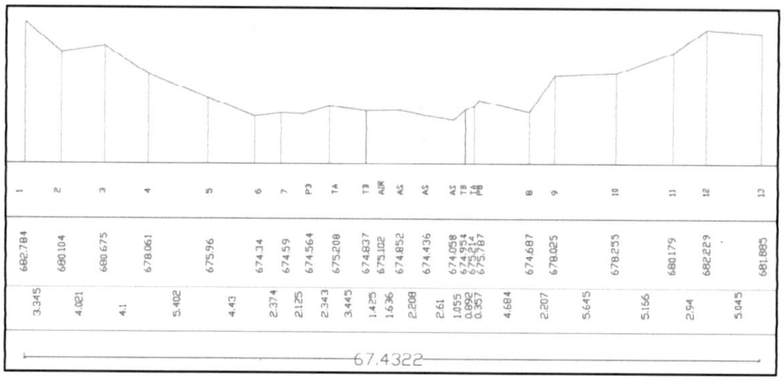

Figure 2. Cimahi River cross section

Based on the cross-section, the maximum capacity discharge of the Cimahi river is 88.28 m^3/s. Cilember River which is located adjacent to the Cimahi River has a cross-section as is shown in Figure 3,

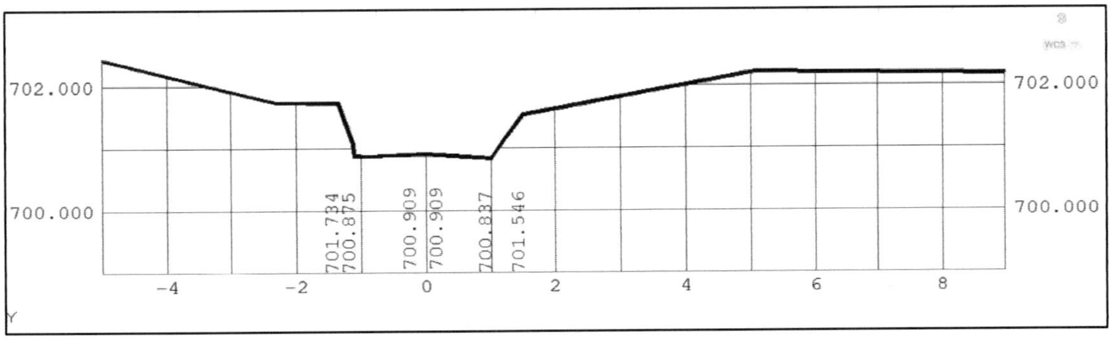

Figure 3. Cilember River cross-section

Cilember river has a very narrow cross-section with a maximum discharge of 9.62 m^3 / s. The Citarik River, which meets the Citarum River is located in the Rancaekek sub-district, the cross-section of the river after being normalized is as is shown in Figure 4,

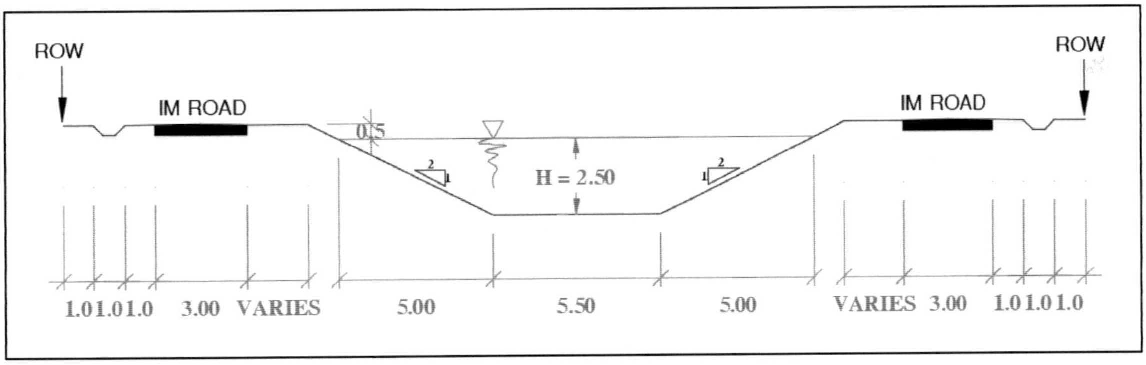

Figure 4. Citarik River cross section

The maximum discharge of the Citarik river in the cross-section is 348.1 m^3/s. The Cisangkuy river which meets the Citarum river in Banjaran sub-district has the cross-section as is shown in Figure 5,

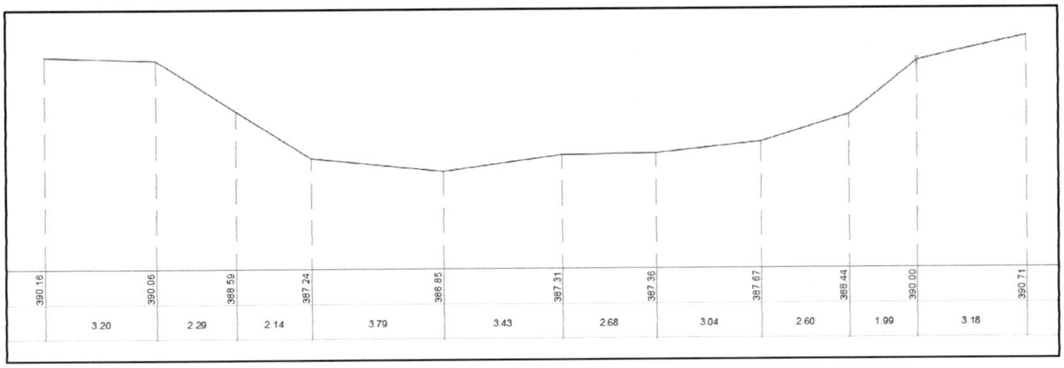

Figure 5. Cisangkuy River cross-section

The maximum discharge of the Cisangkuy river at the cross-section is 328.1 m^3/s.

The unit hydrograph parameters of each river are shown in table 1 below,

Table 1. Parameters of unit hydrograph

Name of The River	Unit Hydrograph Parameter		
	Q_p (m^3/s)	T_p (hour)	T_b (hour)
Cimahi	3.52	2.80	20.00
Cilember	0.75	2.44	25.00
Cisangkuy	5.37	4.00	21.00
Citarik	4.31	3.54	30.00

Based on the above table, it can be seen that the Cilember watershed has a small capacity with a peak discharge of 0.75 m^3 / s. Narrowing the river as much as 60% of its capacity, as well as drainage constraints in the downstream areas at its meeting with the Citarum river, indicate severe river damage.

3.3. The monthly load discharge

Load discharge analysis in this study is made monthly to be able to detect flood events in detail. Discharge is calculated with a return period of 50 years using the convolution method. The monthly discharge for all four rivers can be seen in Table 2 below,

Table 2. Load and capacity discharge

River Name	January	February	March	April	Mei	June	July	August	September	October	November	December
Cimahi												
Qload(m^3/s)	39.98	42.74	40.22	55.87	60.16	32.72	32.07	22.26	36.15	63.51	54.82	48.55
Q_{cap}(m^3/s)	88.28	88.28	88.28	88.28	88.28	88.28	88.28	88.28	88.28	88.28	88.28	88.28
Cilember												
Qload(m^3/s)	25.93	27.73	26.09	36.24	39.02	21.23	20.80	14.44	23.45	41.20	35.56	31.49
Q_{cap}(m^3/s)	9.62	9.62	9.62	9.62	9.62	9.62	9.62	9.62	9.62	9.62	9.62	9.62
Cisangkuy												
Qload(m^3/s)	259.44	351.62	293.70	303.60	264.38	211.50	186.45	162.68	228.52	281.14	386.01	392.80
Q_{cap}(m^3/s)	328.1	328.1	328.1	328.1	328.1	328.1	328.1	328.1	328.1	328.1	328.1	328.1
Citarik												
Qload(m^3/s)	256.58	298.93	399.92	225.87	267.48	213.18	158.93	126.89	151.04	409.10	407.58	397.41
Q_{cap}(m^3/s)	348.1	348.1	348.1	348.1	348.1	348.1	348.1	348.1	348.1	348.1	348.1	348.1

Based on the calculation results of the monthly discharge shown in the above table, there are different flood categories for each river. Cimahi River does not experience floods throughout the year. Cisangkuy and Citarik rivers flooded only in a particular month, while the Cilember river flooded throughout the year. Cilember River flows through a densely populated area in South Cimahi sub-district which is also an industrial area. More than 100 industries are located on the banks of the Cilember river with the collection and disposal of waste into this river. Cisangkuy River is in the Bandung Basin region with its headwater in Pangalengan District at an elevation of 1475 m MSL. The meeting of this river with the Citarum river is in the sub-district of Baleendah which is the heaviest flood area in the Bandung Basin. Citarik River is a second-order river that empties into the Citarum river, Rancaekek sub-district, Bandung Regency at geographical coordinates of 107,94^{0}BT and 6,96^{0}LS at 616.81 m MSL. Citarik River has a length of 51.8 km and its upstream is in South Sumedang, Sumedang Regency at coordinates 107.7^{0}BT and 6,99^{0}LS at an elevation of 1210 m MSL. This watershed area is 332 km^2 with an average slope of 8.7%. The flood inundation area is located at an average slope of 1.1%, in the area where the river meets the Citarum river. The inundation area is approximately 50% of the area of the Rancaekek sub-district and 7.2% of the area of the Citarik River Basin.

This category can be quantified based on the index of watershed damage, I_D. The results are shown in Table 3,

Table 3. The index of watershed damage (I_D)

River Name	January	February	March	April	Mei	June	July	August	September	October	November	December
Cimahi												
Q_{load} (m^3/s)	39.98	42.74	40.22	55.87	60.16	32.72	32.07	22.26	36.15	63.51	54.82	48.55
Q_{cap} (m^3/s)	88.28	88.28	88.28	88.28	88.28	88.28	88.28	88.28	88.28	88.28	88.28	88.28
Q_{load}/Q_{cap}	0.45	0.48	0.46	0.63	0.68	0.37	0.36	0.25	0.41	0.72	0.62	0.55
I_D	0.00	0.00	0.00	0.00	0.00	0.00	0.00	0.00	0.00	0.00	0.00	0.00
Level of damage	no damage	no damage	no damage	no damage	no damage	no damage	no damage	no damage	no damage	no damage	no damage	no damage
Cilember												
Q_{load} (m^3/s)	25.93	27.73	26.09	36.24	39.02	21.23	20.80	14.44	23.45	41.20	35.56	31.49
Q_{cap} (m^3/s)	9.62	9.62	9.62	9.62	9.62	9.62	9.62	9.62	9.62	9.62	9.62	9.62
Q_{load}/Q_{cap}	2.70	2.88	2.71	3.77	4.06	2.21	2.16	1.50	2.44	4.28	3.70	3.27
I_D	1.70	1.88	1.71	2.77	3.06	1.21	1.16	0.50	1.44	3.28	2.70	2.27
Level of damage	severe	severe	severe	severe	severe	severe	severe	moderate	severe	severe	severe	severe
Cisangkuy												
Q_{load} (m^3/s)	259.44	351.62	293.70	303.60	264.38	211.50	186.45	162.68	228.52	281.14	386.01	392.80
Q_{cap} (m^3/s)	328.10	328.10	328.10	328.10	328.10	328.10	328.10	328.10	328.10	328.10	328.10	328.10
Q_{load}/Q_{cap}	0.79	1.07	0.90	0.93	0.81	0.64	0.57	0.50	0.70	0.86	1.18	1.20
I_D	0.00	0.07	0.00	0.00	0.00	0.00	0.00	0.00	0.00	0.00	0.18	0.20
Level of damage	no damage	low	no damage	no damage	no damage	no damage	no damage	no damage	no damage	no damage	low	moderate
Citarik												
Q_{load} (m^3/s)	256.58	298.93	399.92	225.87	267.48	213.18	158.93	126.89	151.04	409.10	407.58	397.41
Q_{cap} (m^3/s)	348.10	348.10	348.10	348.10	348.10	348.10	348.10	348.10	348.10	348.10	348.10	348.10
Q_{load}/Q_{cap}	0.74	0.86	1.15	0.65	0.77	0.61	0.46	0.36	0.43	1.18	1.17	1.14
I_D	0.00	0.00	0.15	0.00	0.00	0.00	0.00	0.00	0.00	0.18	0.17	0.14
Level of damage	no damage	no damage	low	no damage	no damage	no damage	no damage	no damage	no damage	low	low	low

Based on the results of the above table, the Cimahi river area is not damaged. Citarik River and Cisangkuy River are categorized as low damaged and the Cilember River is heavily damaged.

4. Conclusion

Quantification of watershed damage is needed to quantitatively measure the level of watershed damage. With this quantification value, monitoring changes in watersheds can be carried out so that watershed damage can be determined earlier and helping scale priorities for river basin improvement as a guideline for stakeholders. Noteworthy is the validity of the quantification formula so that it approaches its actual condition.

5. References

[1] Fauzan Ali. 2019. Fragrant Citarum and Hope for Watershed Recovery. LIPI Legal Bureau (Citarum Harum dan Harapan Pemulihan Daerah Aliran Sungai. Biro Hukum LIPI)

[2] Dewi Nur Hayati. 2019. The Citarum River Basin is a Critical Problem in Indonesia. Upstream Citarum Watershed Management Program Synchronization Workshop (DAS Citarum Adalah Masalah Kritis di Indonesia. Lokakarya Sinkronisasi Program Pengelolaan SDA-LH DAS Citarum Hulu) , Universitas Padjadjaran.

[3] Ariani BS & Ramli. 2015. Analysis of River Network Integration and Drainage Systems as a Cimahi City Flood Solution (Analisis Keterpaduan Jaringan Sungai dan Sistem Drainase Sebagai Solusi Banjir Kota Cimahi). SNIJA Proceeding 2015, 73-78

[4] Hundecha, Y., & Andras, B. 2009. Modeling of The Effect of Land Use Change on The Runoff Generation of a River Basin through Parameter Regionalization of a Watershed model. Journal of Hydrology, 281-295.

[5] Fortunata U.Msoffe, M. Y. 2011. Spatial Correlates of Land-use Changes in The Maasai-Steppe of Tanzania: Implications for conservation and environmental planning. International Journal of Biodiversity and Conservation, 280-290

[6] Sevim Pelin Ozkan & Cigdem Tarhan. 2015. Detection of Flood Hazard in Urban Area Using GIS Izmir Case. Procedia Technology 22, 373-381

[7] M.Syahril Badri Kusuma, Harkunti P.Rahayu, Muhammad Farid, M.Bagus Adityawan, Tia Setiawati, Rasmiaditya Silasari. (2010). Study of Development of Flood Risk Index Map in Bukit Duri Village, Jakarta (Studi Pengembangan Peta Indeks Risiko Banjir pada Kelurahan Bukit Duri Jakarta). Jurnal Teknik Sipil V.17 No 2, 123-134

[8] Edwin, Amrizal Saidi, Apprisal, Yulnafatmawita, Ita Carolita. 2015. Spatial and Temporal Analysis of Land Use Change for 11 Years (2004 – 2014) in Sub-Watershed Sumpur Singkarak. International Journal on Advanced Science, Engineering and Information Technology (IJASEIT) V.5 No 5: 326 -329

[9] I Gede Tunas and Rizaldi Maadji. 2018. The Use of GIS and Hydrodynamic Model for Performance Evaluation of Flood Control Structure. International Journal on Advanced Science, Engineering and Information Technology (IJASEIT) V.8 No 6: 2413 -2420

[10] Mahmud, Ambar Kusumandari, Sudarmadji, Nunuk Supriyatno. (2018). A Study of Flood Causal Priority in Arui Watershed, Manokwari Regency, Indonesia. Jurnal Manajemen Hutan Tropika V 24 No 2: 81 - 94

[11] Avinash, S. 2014. Flood Related Disasters: Concerned to Urban Flooding in Bangladesh India. Journal of Spatial Hydrology, 76-83.

[12] Edsel B Daniel, J. V. (2010). Watershed Modelling Using GIS Technology: A Critical Review. Journal of Spatial Hydrology, 13-23.

[13] Anwar, M. R. (2011). The Rainfall-Runoff Model Using of The Watershed Physical Characteristics Approach. Journal of Civil and Environmental Engineering, 71-75.

[14] Ariani Budi Safarina, Iin Karnisah, Ade Sena Permana, Dadang Sufianto, 2019. Green Cimahi Watershed for Balancing Water Supply and Flood Control. IOP Conf.Series: Materials Science and Engineering 669, 012040, pp. 1-11

[15] Ariani Budi Safarina, Iin Karnisah, Ade Sena Permana. 2019. Study of The Effect of River Basin Morphology Change on Threshold Parameters in Cimahi Flood Early Warning System. IOP Conf.Series: Earth and Environmental Science 273 012013, pp. 1-8

[16] Ariani Budi Safarina, Ramli, Muhammad Shiddiq Sayyid Hasharon. 2017. Two Alert Flood Early Warning System Method Based on Rainfall-Runoff Model International Journal of Civil Engineering and Technology 8(12): 1032-1044

Acknowledgments

This research is fully supported by UNJANI Competitive Research SKEP/129/UNJANI/V/2020. The authors fully acknowledged Jenderal Achmad Yani University for the approved fund which makes this important research viable and effective.

Identification of paddy fields using temporal NDVI changes in Minamiaso Village, Kumamoto Prefecture, Japan to estimate groundwater recharge

H Amano[1], Y Iwasaki[2] and T Ichikawa[1]

1 Liberal Arts Education Center, Kumamoto Campus, Tokai University, 9-1-1 Toroku, Higashi-ku, Kumamoto 862-8652, Japan
2 Department of Electrical Engineering and Computer Science, Faculty of Industrial and Welfare Engineering, Tokai University, 9-1-1 Toroku, Higashi-ku, Kumamoto 862-8652, Japan
E-mail: ah069881@tsc.u-tokai.ac.jp

Abstract. In Minamiaso Village (Kumamoto Prefecture, Japan), land covers of forests, grasslands, and paddy fields play an important role in groundwater recharge. The land covers have been changed in a past few decades by agricultural systems and natural disasters: landslides, slope failures, and damage of irrigation canals caused by the 2016 Kumamoto earthquake in Japan. In this study, we extracted training data for classifying the land covers, and detected the temporal change of NDVI (Normalized Difference Vegetation Index) for each land cover. To quantify the impact of past land cover change on the groundwater recharge, we need to grasp the area of each land cover with the different hydrological parameters such as infiltration capacities. In the case of paddy fields, its distribution also should be known because the infiltration rates that we investigated vary depending on the paddy fields. We attempted to identify the paddy field distribution by the obtained NDVI values, and confirmed that the paddy field areas near the earthquake faults damaged by the earthquake where are still not usable were well distinguished. The paddy field distribution based on NDVI was almost matched with that based on SAR.

1. Introduction

Aso Caldera in Kumamoto Prefecture, located in the center of Kyushu Island in Japan (figure 1), is the source of six rivers in Kyushu Island, and plays a major role in the water cycle in Kyushu Island. Aso Caldera stores a rich groundwater, and many studies on the groundwater and the water cycle system have been done [1-3]. Minamiaso Village whose area is surrounded with the blue dotted line in figure 1 is one of the three municipalities in Aso Caldera, and is located in the south-west of the caldera. The groundwater recharged at forests and grasslands springs out from multiple large and small spring sources. The quantities of spring water from Shirakawa Spring Source and Takezaki Spring Source reach 60 tons and 120 tons per minute, respectively [2]. Therefore, paddy field agriculture using spring water and pumped groundwater is actively practiced in Minamiaso Village. Flooded paddy field is a one of the important groundwater recharge area in addition to forest and grassland. The forests, grasslands, and paddy fields occupy a large part of the area in Mianamiaso Village so that it is recognized that the maintenance of these land covers is essential for preservation of groundwater.

The grassland, which has larger groundwater recharge rate compared to the forest [4], has been maintained by artificial disturbance such as Noyaki (open burning) and mowing. Noyaki (open burning) means burning the grasslands to prevent shrub growth and the transition to forests, and to exterminate

IOP Conf. Series: Earth and Environmental Science **633** (2021) 012012 doi:10.1088/1755-1315/633/1/012012

noxious insects [5]. However, a large part of grasslands has been transitioning to forests mainly due to the decrease and ageing of farmers who are managers. On the other hands, the paddy fields significantly decreased because of policy of reducing rice acreage. Moreover, the Kumamoto earthquake (main shock) occurred on April 16, 2016 gave serious damages such as the reduction of grassland and forest areas due to landslide disasters, and the suspension of paddy field agriculture due to the damage of irrigation canals. To quantify the impact of past land cover changes on the groundwater recharge, we need to grasp the area of each land cover with the different hydrological parameters such as infiltration capacities. In the case of paddy fields, its distribution also should be known because the infiltration rates that we investigated vary depending on the paddy fields as shown in figure 1.

To identify and classify the land covers, satellite remote sensing is effective. Classification of land covers and the evaluation for the transition of land covers using NDVI (Normalized Difference Vegetation Index) have been done [6-9]. We have detected the locations of landslide disasters in Aso using NDVI values at two periods before and after the earthquake [10]. Our final goal is to grasp the land covers change in a past few decades including recent natural disaster using BDT (Binary Decision Tree) based on NDVI. In this study, we extract training data, and detect the temporal change of NDVI for each land cover. To measure accurately the temporal change of NDVI for each land cover, we measure in detail the temporal change of NDVI for each crop specie in paddy fields and upland fields, and for each plant in grasslands. Then, we attempt to identify the paddy field distribution by the obtained NDVI values, and confirm the paddy field areas near the earthquake faults damaged by the earthquake. We also compare the paddy field distribution based on NDVI with that based on SAR.

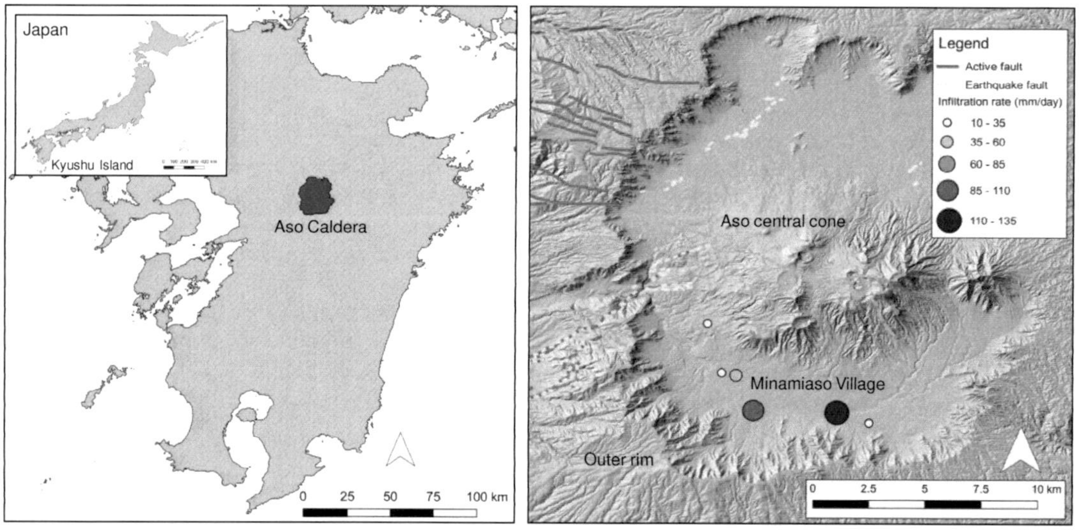

Figure 1. The locations of Aso Caldera and Minamiaso Village, and infiltration rates of paddy fields.

2. Materials and methods

2.1. Location of each land cover

In order to evaluate the temporal change of NDVI for each land cover, it is first necessary to obtain the location of each land cover. As for agricultural crops, we selected the species with a cultivation area of 10 ha or more based on 2016 account book of paddy field for Minamiaso village. We calculated the area of each species for each village division, Oaza, selected the five fields with larger areas so as to exceed 10% of the total area, and identified the positions of selected fields using "Agriculture land information system agriculture committee" [11]. As for damaged paddy fields, because it is unclear how the damaged paddy fields will be used after the earthquake, we analyzed all damaged paddy fields of 900 m^2 or more to understand the average NDVI characteristic. Forests (conifers and hardwoods), grasslands, lawns, and bare lands were specified from "1/25,000 vegetation survey (6th and 7th)" [12]. We specified buildings such as elementary and junior high schools, government offices with large site areas, and temporary houses built after the earthquake from Google maps satellite view. Table 1 shows the details of the classification of land covers specified in the analysis.

Table 1. Classification of detail land covers and data sources.

Large category	Medium category		Small category		Data source
	ID	Land cover	ID	Land cover	
Paddy field	P	Paddy field	P1	Akigeshiki rice	Account book of paddy field
			P2	Koshihikari rice	
			P3	Hitomebore rice	
			P4	Hinohikari rice	
			P5	Morinokumasan rice	
			P6	WCS	
			P7	WCS - Italian ryegrass	
Damaged paddy field	DP	Damaged paddy field	DP	Damaged paddy field	
Upland field	U	Upland field	U1	Italian ryegrass	
			U2	Aoba millet	
			U3	Aoba millet - Italian ryegrass	
			U4	Italian ryegrass - Italian ryegrass	
			U5	Natsu kansou (Setaria)	
			U6	Buckwheat	
			U7	Soybean	
Conifer	C	Conifer	C	Conifer	1/25,000 vegetation map
Hardwood	H	Hardwood	H	Hardwood	
Grassland	G	Grassland	G1	Japanese pampas grass	
			G2	Turf	
			G3	Bamboo grass - Japanese pampas grass	
			G4	Pasture	
	L	Lawn	L	Lawn for golf	
Building	B	Building	B	Building (House, School, Town office)	Google maps satellite view
	PG	Plastic Greenhouse	PG1	Tomato	Account book of paddy field
			PG2	Asparagus	
Bare land	BL	Bare land	BL	Bare land	1/25,000 vegetation map

2.2. Preprocesses and analysis of satellite images

To detect temporal changes of NDVI for areas of classified land covers, we selected the Landsat-8 multi-spectral images of five periods with extremely few clouds of year 2016 on January 16, March 23, May 23, August 11, and October 20. These images were downloaded from LandsatLook Viewer [13]. In Aso caldera, it is too difficult to obtain Landsat-8 images with no clouds in the higher time resolution because Aso caldera is easily covered by clouds due to the geographical feature. Landsat-8 has Operation Line Imager (OLI) and Thermal Infrared Sensor (TIRS), and acquires 11 bands with different wavelengths. NDVI is calculated by equation (1) from the band 4 (visible red band) and band 5 (near-infrared band).

$$\mathrm{NDVI} = (NIR - R) / (NIR + R), \quad (1)$$

where *NIR* is the reflectance of the near-infrared band, and *R* is the reflectance of visible red band. NDVI outputs the values between -1.0 and 1.0 which represents the activity level of vegetation. We calculated the mean values of NDVI in the polygons of determined locations, and evaluated temporal change of NDVI for each land cover. We also attempted to extract paddy fields based on the obtained NDVI values. Moreover, we attempted to extract paddy fields using two SAR (Synthetic Aperture Radar) data at different periods [14]. We used the C-band vv polarization Sentinel-1 SAR images on May 26, and August 30, 2016, which were downloaded from Copernicus Open Access Hub [15]. The rice planting season in Minamiaso Village is May. Therefore, the image on May 26, 2016 shows flooded paddy fields. In flooded paddy fields, the irradiated microwaves are specularly reflected, so that the backscattering is small and they are displayed in black. However, water areas such as rivers in the image are also displayed in black. As an image to mask such areas, we used the image on August 30, 2016 when the rice grew and the backscattering increased. When the difference between the SAR images at the two periods is calculated, it is considered that the gray levels in the paddy fields where the backscattering greatly changes are higher than those of other land covers. So, the regions with high gray levels in the difference image were extracted as paddy fields.

3. Results and discussions

3.1. Temporal changes of NDVI for small category

Figure 2 shows obtained temporal change of NDVI for each land cover. The trends of the temporal changes of NDVI for paddy fields shown in figure 2a are similar each other regardless of the crop species. NDVI values decrease from March to May due to rice planting after puddling in May. On the other hand, we consider that NDVI value increases due to weed breeding because damaged paddy fields (DP) are to be left without puddling. Italian ryegrasses (P7), secondary crops in paddy fields, are seeded in October, and are harvested from mid-April to early May. Thus, we consider that NDVI value from January to March of Italian ryegrasses (P7) is higher than other crop species because Italian ryegrasses grow from winter to spring.

The trends of the temporal changes of NDVI for upland fields shown in figure 2b are almost similar each other except for Italian ryegrasses (U1), buckwheat (U6), and soybeans (U7). This is because the cultivation time differs depending on the crop species. The NDVI values of forage crops are generally higher than those of buckwheat (U6) and soybeans (U7). This may be because forage crops regenerate after harvesting.

The NDVI values of forests (conifers (C) and hardwoods (H)) shown in figure 2c increase from January to August, and decrease in October, which correspond to seasonal changes. In the grasslands of Aso Caldera, Noyaki (open burning) is performed in early March. Thus, unlike forests, the NDVI values of grasslands do not increase from January to March. The NDVI values of turf (L) in golf courses show a decrease in August, which are different from the NDVI values of paddy fields, upland fields, forests (conifers and hardwood), and grasslands. The cause is not clear, but is presumed to be related to the type of turf used in the golf courses.

The NDVI values of buildings (B) shown in figure 2d indicate a peak in May. The trend of the temporal change of NDVI of buildings (B) is similar to that of turf in the golf courses. The harvest time of tomatoes is from July to September, that of asparagus is in May, and the two harvest times are different. So, we predicted that the temporal changes of NDVI of tomatoes and asparagus are also different. However, they (PG1 and PG2) show almost the same changes. Because the spatial resolution of the satellite images is 30 m, it may represent the characteristics of the surrounding ridgeways and upland fields rather than the plastic greenhouses. The NDVI values of bare lands are almost no change, and are close to zero.

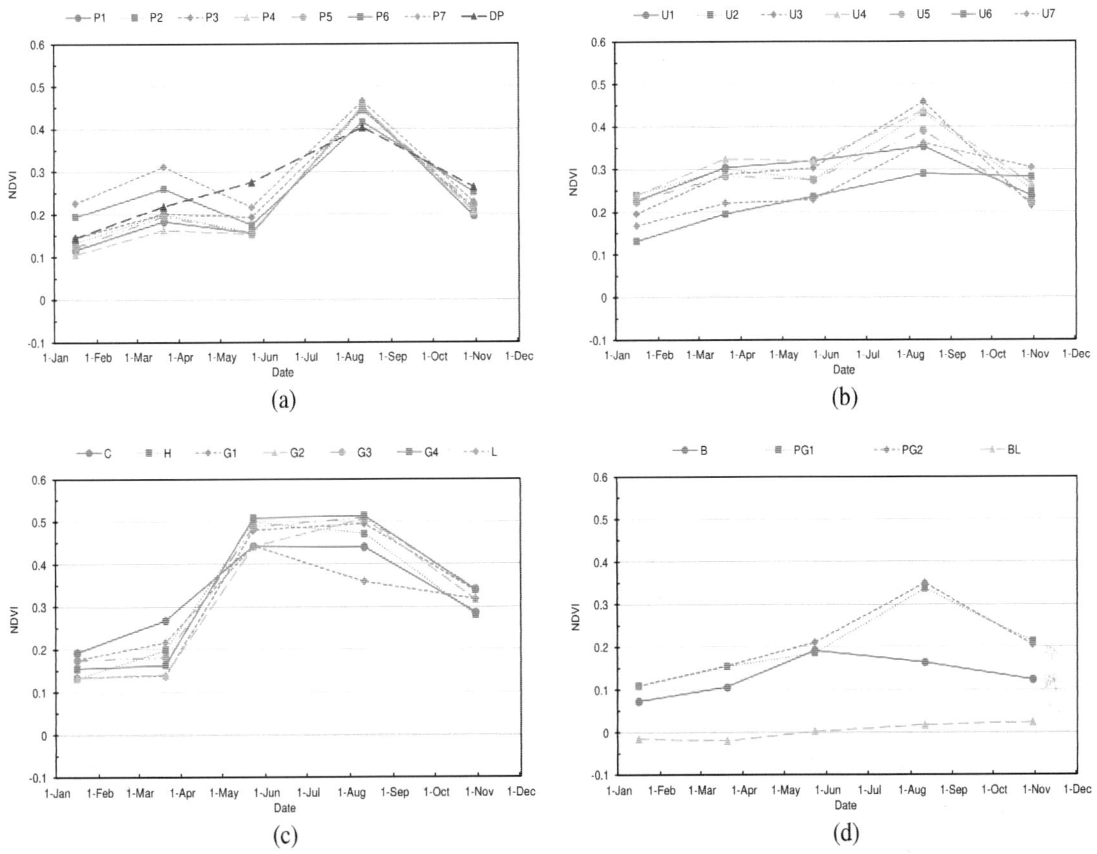

Figure 2. Temporal changes of NDVI based on small category: (a) paddy field and damaged paddy field; (b) upland field; (c) conifer, hardwood, grassland and lawn; (d) building, plastic greenhouse and bare land.

3.2. Temporal changes of NDVI for medium category

Figure 3 shows the temporal changes of NDVI for the ten medium categories shown in table 1, which are the averages of those for the small categories shown in figure 2. The NDVI values of paddy fields of rice planting time on May are lower than the others. Thus, the areas where NDVI values on May are less than 0.2174 are extracted first. The value 0.2174 is the mean value of P7 (WSC- Italian ryegrass) with the highest NDVI value for paddy fields on May. However, by using this threshold value, the areas of bare lands, plastic greenhouses, and buildings are also extracted. Next, by using the NDVI values on August, these three areas are separated from paddy fields. The threshold value is 0.3414 of P6 (WCS) with the lowest NDVI value (mean value minus standard deviation) for paddy fields on August. However, the area above 600 m a. s. l. on Aso central cone and above 500 m a. s. l. on outer rim were excluded.

Figure 4 shows the analysis results. Figure 4a shows the paddy fields in JAXA (Japan Aerospace Exploration Agency) high-resolution land use and land cover map of Japan [16] which provides the average situation from 2006 to 2011 with a resolution of 10 m before the earthquake. Figure 4b shows the distribution of paddy fields obtained by the method described above. Figure 4c shows the distribution of paddy fields identified based on SAR which is described in Section 2.2. Figure 4d is a superposition of figure 4b,c for comparison. The extracted area of paddy field in figure 4b is 11,325,898 m^2. This area generally corresponds to the area of 11,999,675 m^2 reported by account book of paddy field. Comparing figure 4a and 4b, it is obvious that the paddy fields decreased. Especially, the areas near the earthquake faults shown in yellow lines [17] in figure 4, which were confirmed to have appeared on the ground surface when the earthquake occurred, are still damaged, and still cannot be used as paddy fields. Figure 4b,c seems to represent the situation well. The distributions of paddy fields in figure 4b,c are almost the same, but the area seems to be smaller in figure 4b. Because the NDVI threshold values and the order of extraction for land covers affect the areas of extracted paddy fields, we will consider these points in a future work.

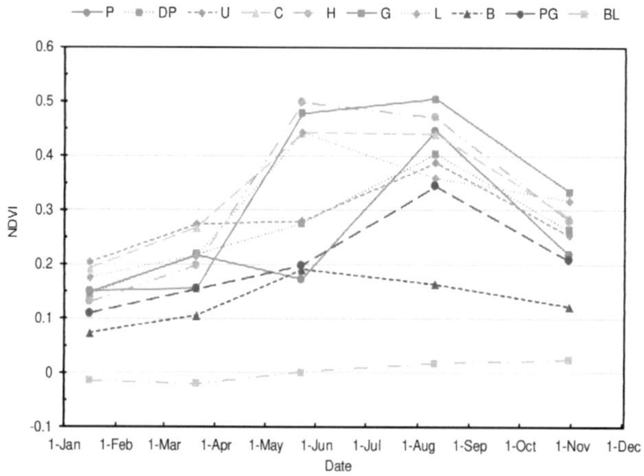

Figure 3. Temporal changes of NDVI based on medium category.

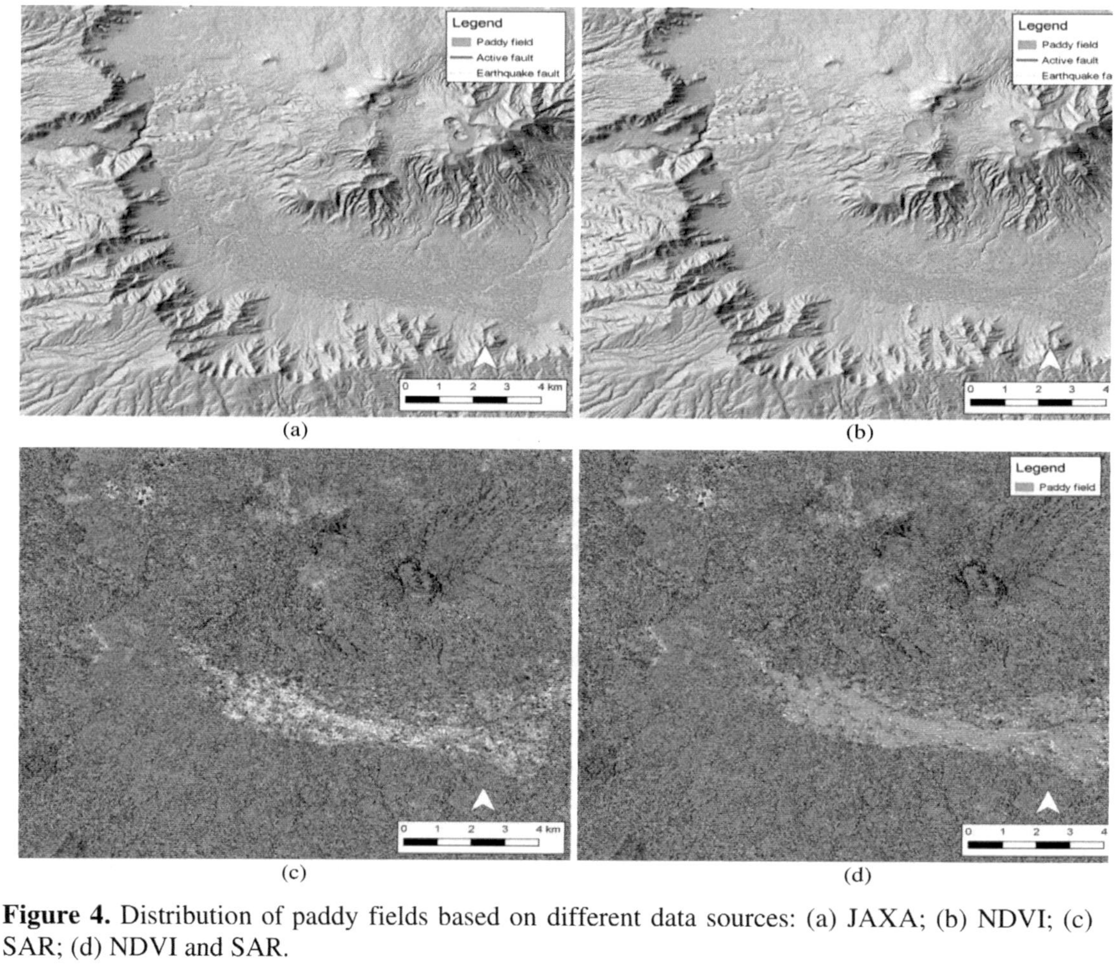

Figure 4. Distribution of paddy fields based on different data sources: (a) JAXA; (b) NDVI; (c) SAR; (d) NDVI and SAR.

4. Conclusions

In this study, we extracted training data, and investigated the temporal changes of NDVI. The obtained NDVI values showed the characteristics changed depending on land covers and season. For example, the NDVI values are low in paddy fields in rice planting time and in grasslands in Noyaki (open burning) season. Then, we attempted to identify the paddy field distribution by the obtained NDVI values, and confirmed that the paddy field areas near the earthquake faults damaged by the earthquake where were still not usable were well distinguished. The extracted area of paddy field generally corresponds to the area reported by account book of paddy field. The distribution of paddy fields based on NDVI is generally good agreement with that based on SAR. The purpose of identifying the paddy field distribution is to estimate the groundwater recharge in the area.

We plan to identify the distribution of other land covers, such as grasslands, forests (conifers and hardwoods), and upland fields, using the temporal change of NDVI for each land cover obtained in this study as training data. Then, we will quantify the impact of past land cover changes on the groundwater recharge.

5. References

[1] Shimotsu M 1986 On the characteristics of groundwater in the hydrologic cycle of the Aso Volcano *J. Jpn. Assoc. Groundwater Hydrol.* **28** 1-13 (in Japanese with English abstract)

[2] Shimano Y 1997 Hydro-chemical study of spring waters in the Aso caldera, central Kyushu *Bunsei bulletin* **8** 43-67 (in Japanese)

[3] Kagabu M, Shimada J, Shimano Y, Higuchi S and Noda S 2011 Groundwater flow system in Aso caldera *J. Jpn. Assoc. Hydrolog. Sci.* **41** 1-17 (in Japanese with English abstract)

[4] Kudo K, Shimada J, Maruyama A and Tanaka N 2016 The quantitative evaluation of groundwater recharge rate using Displacement Flow Model with stable isotope ratio in the soil water of difference vegetation cover *Jpn. Groundwater Hydrol.* **58(1)** 31-45 (in Japanese with English abstract)

[5] Yasunaka Y, Oishi K, Anzai H, Miwa M, Kumagai H, Hirooka H and Ieiri S 2015 Asseing the effects of controlled burning and grazing on vegetation change in the grasslands of Aso region using satellite images analyses *J. Jpn. Agric. Syst. Soc.* **31** 117-125 (in Japanese with English abstract)

[6] Lunetta R S, Knight J F, Ediriwickrema J, Lyon J and Dorsey Worthy L 2006 Land-cover change detection using multi-temporal MODIS NDVI data *Remote Sens. Environ.* **105** 142-154

[7] Jia K, Liang S, Wei X, Yao Y, Su Y, Jiang B and Wang X 2014 Land cover classification of Landsat data with phenological features extracted from time series MODIS NDVI data *Remote Sens.* **6** 11518-11532

[8] Li Y, Cao Z, Long H, Liu Y and Li W 2017 Dynamic analysis of ecological environment combined with land cover and NDVI changes and implications for sustainable urban-rural development: The case of Mu Us Sandy Land, China *J. Clean. Prod.* **142** 697-715

[9] Han J-C, Huang Y, Zhang H and Wu X 2019 Characterization of elevation and land cover dependent trends of NDVI variations in the Hexi region, northwest China *J. Environ. Manage.* **232** 1037-1048

[10] Iwasaki Y, Tamaki T, Murata K, Koga A and Fujimoto K 2020 Detection of land cover changes before and after the 2016 Kumamoto earthquake in Japan using remote sensing for evaluation of environmental impacts. *Proceedings of 2020 10th International Conference on Future Environment and Energy (ICFEE 2020)* Kyoto Japan 7-9 January 2020 184-191.

[11] Agriculture land information system agriculture committee. Available online: https://www.alis-ac.jp/ (accessed on 28 January 2020).

[12] Biodiversity center of Japan. 1/25,000 vegetation survey (6th). Available online: http://gis.biodic.go.jp/webgis/sc-006.html (accessed on 31 January 2020).

[13] U.S. Geological Survey. LandsatLook Viewer. Available online: https://landsatlook.usgs.gov/ (accessed on 17 December 2019).

[14] Kimura A and Shimamura H 2015 Classification of planting condition of paddy fields utilizing two time-series acquistions of high-resolution SAR image *J. Jpn. Soc. Photogramm. Remote Sens.* **54** 118-132 (in Japanese with English abstract)
[15] European Space Agency. Copernicus Open Access Hub. https://scihub.copernicus.eu/ (accessed on 15 February 2020).
[16] Japan Aerospace Exploration Agency. High-Resolution Land Use and Land Cover Map of Japan (Released in September 2016/ Version 16.09). Available online: https://www.eorc.jaxa.jp/ALOS/en/lulc/lulc_index.htm (accessed on 17 February 2020).
[17] Suzuki Y, Ishimura D, Kumaki Y, Kumahara Y, Chida N, Nakata T and Nakano Y 2017 1:25,000 Active fault map Aso *Geospatial Information Authority of Japan*

Acknowledgments
This research was performed by the Environment Research and Technology Development Fund (JPMEERF19S20505) of the Environmental Restoration and Conservation Agency of Japan.

IOP Conf. Series: Earth and Environmental Science **633** (2021) 012013 doi:10.1088/1755-1315/633/1/012013

Non-destructive measurement of rice leaf dimension in 3D point cloud

P Rattanasuwan[1], I Kumazawa[1] and T Kasetkasem[2]

1 Department of Information and Communication Engineering, Tokyo Institute of Technology, Yokohama 226-0026, Japan
2 Department of Electrical Engineering, Kasetsart University, Bangkok 10900, Thailand
E-mail: rattanasuwan.p.aa@m.titech.ac.jp

Abstract. In order to try to improve the yield, knowing the expanding of a leaf by measuring its dimension can be one easy way to monitor the plant growth. While rice is considered to be one of the most important agricultural plant, their leaves are quite difficult to measure the size since they are mostly small and direct measurement could harm the plant. To do so, we deploy the non-destructive system to measure the dimension of the rice leaf from a point cloud. The proposed system takes only photos from a smartphone camera to produce the 3D model of a specific leaf using structure from motion and multi-view stereo. To measure the dimension of the rice leaf in metric units, the scale reference is placed into the scene before the photo taking process to involve it in the reconstruction result. From the reconstructed point cloud, the leaf veins can be extracted in order to find the accurate width of a leaf based on thresholding technique using combination of RGB and HSV spaces. By comparing with measured length of the leaf in point cloud with length of one block in scale reference, we can eventually obtain the distant between each leaf vein in the metric system unit.

1. Introduction

The fact that rice is one of the most consumed plant is not uncommon knowledge. For covering the demand of people in consuming, the improvement of yield is necessary. To do so, the monitoring of rice growth can be a factor to help understanding the best environment to grow rice and may can also help reducing the cost of growing. Despite the attempt to find the optimal environment for rice, but without the proper plant growth measurement, it is difficult to determine which is the best environment. There are many ways to perform the growth monitoring, the simplest one would be to see the expanding of a leaf. Generally, it can be done by directly measure a leaf from a tree with a simple measurement tool such as a ruler. However, with this approach, there is a possibility of a tree to get damaged since in some case the leaf may have to be torn or taken out. For example, Maloof et al. has proposed a method [1] that can accurately measure the leaves parameters but the leaves have to be taken out of the tree in order to be measured. As a result, the new approach of measuring a leaf without direct contacting and non-destructive has to be proposed.

The measurement from using of 2D or 3D images is not a new thing. For several past decades, the leaf measurement systems using images have been constantly published. Obliviously the using of 2D images is far simpler and more straight forward. However, in many cases, the results are heavily depending on relative position of the camera with respect to the plant which lead to unsatisfied result when applying to other environment, for instance, the method proposed by Easlon and Bloom [2] which takes 2D image from top view of a plant along with the calibration areas can accurately

measure under the only condition that the calibration area is kept in the same plane as the leaves in order to avoid perspective distortion. On the other hand, 3D system is far more accurate but usually come up with higher costs in term of money and computation. Thus, we would like to create a simple system that can measure a leaf without direct contact by using the 3D reconstruction technique.
There are variety of 3D reconstruction methods, some use a ToF camera [3] or LiDAR sensor [4] while some use a laser scanning system [5]. But most widely used one is from a normal RGB camera since it can apply to many applications while keeping the cost low. The fundamental yet effective technique of 3D reconstruction is Structure from Motion (SfM) [6]. SfM reconstructs 3D model from two or more images using the concept of stereo vision. At the present, there are many programming libraries and software providing SfM function for free. Even though it can reconstruct 3D model effectively, SfM alone can only reconstruct a sparse model, which means it cannot reconstruct fine detail of an object surface. With the aid of Multi view stereo (MVS) [7], the sparse reconstruction result from SfM can be more detail which called a dense reconstruction. There are numbers of MVS based technique, one of the most effective is Pixel-wise Multi view stereo (PMVS) [8] which take pixel information from corresponding 2D images to help enhance the detail of the model surface. With both techniques, the 3D model of an object can finally be generated with a fine detail.
The objective of this research is to monitor the growth of a rice plant by measuring the expanding of a leaf with 3D measurement technique to see how much it grows in a period of time. Although there are many ways to acquire the 3D model of a rice leaf, in order to make the system compact, we chose the camera from a smart phone to use in the 3D reconstruction process. The SfM and MVS have been used to reconstruct the 3D model of a rice leaf. For evaluation of the leaf expanding, the distance between each leaf veins are estimated from the generated point cloud. To do so, the determining of which parts of point cloud are leaf veins or leaf surfaces is essential. Hence, the process of how to classify the leaf veins and the measurement of a distance between each leaf vein in a metric system is described in this paper.

2. Methodology

This section describes the theories involved in this research and the process lead to the result of distance between leaf veins. An overview of the proposed system is illustrated in figure 1. The proposed system starts with the collection of the 2D image dataset of a rice leaf. From a set of 2D images, SfM and MVS are used to generate the 3D point cloud, then use the generated point cloud to measure the distance between each leaf vein by choosing one leaf from the whole model, select points that stay in the same line of veins, determine whether each point is leaf vein or leaf surface with the thresholding of RGB and HSV, then measure the distance in a metric system respectively. The detail of each stage is described in this section.

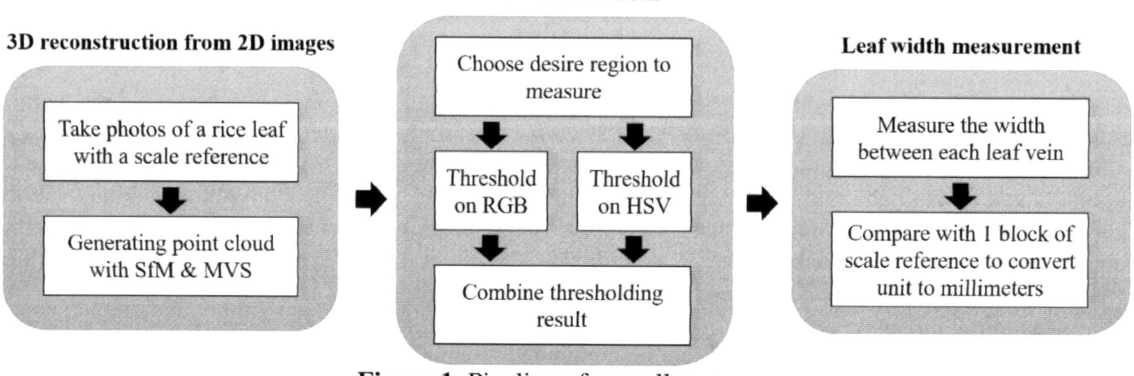

Figure 1. Pipeline of overall system

2.1. 3D reconstruction from 2D images

In order to reconstruct the 3D point cloud from an object, a number of photos of an object is required. The most convenient way according to our past experiments is to take a video around the desired leaf in the most various angle possible and then extract the video frames to get the photos of the leaf. With this approach, the reconstruction result tends to be more accurate since the feature matching in SfM needs sufficient number of images and angles to accurately estimate the position of features in 3D space. Please note that when taking a video of a leaf, the distant between the leaf and camera should be enough for the camera focus to be exactly on the leaf and the amount of light must be enough to clearly see the leaf detail, i.e. leaf veins. In addition, the extracted photos should be at least 100 images in order to obtain fine reconstruction result. However, if only a leaf is the only object we capture a video of, the measurement of it alone would not be able to tell its size in the real-world unit, therefore, a paper with the chessboard pattern with 1 by 1 centimetre for each block, refer as a scale reference, is put beside the leaf before a video capturing to involve this scale reference in the reconstructed point cloud for further unit converting propose. Please note that the scale reference does not have to be in the chessboard pattern, it could be in any shape but it must contain a specific length we want to use as a reference for the real-world scale, e.g. 5 millimetres.

As described earlier, the 3D reconstruction to obtain a point cloud is performed by the SfM and MVS. While a number of open and free software are available with a variety of algorithms to be used, the authors chose COLMAP [8, 9] as it has a very clean and easy to use interface and also can customize in many ways. COLMAP can perform a sparse reconstruction by SfM and dense reconstruction by MVS. It also can create a mesh of an object with the triangulation of either Poisson or Delauney but only the dense reconstruction from MVS in ply file is enough for our study. Although COLMAP is selected for this work, any other approaches or software that can generate the fine enough point cloud, such as CMVS [10] or using of laser scanner or LiDAR, can also applicable instead. The using of COLMAP essentially gives a point cloud of a specific leaf the same as the one that has been taken photos of.

2.2. Leaf vein extraction and leaf width measurement

After the obtaining of a point cloud, the process of width measurement can be easily done by selecting coordinates of the leaf edge and the right edge of a leaf and then find distance between each two points. However, rice leaves usually have a slight curvature and directly find distant from left to right could lead to inaccurate result. Instead, we find a width between each leaf vein as they are aligned along the curvature of a leaf and such measuring could deal with this problem.

The next problem is how to distinguish between the leaf surface and leaf vein points in point cloud. Basically, it can be distinguished by the colour where the vein is in slightly brighter green colour than the surface. As a result, we extract the vein points by performing the thresholding using the RGB information. But as sometimes, thresholding on the RGB information alone cannot fulfil its purpose, the additionally use of the HSV information can help the system distinguish the vein from a leaf more accurate as HSV can tell the brightness and vividness of a specific colour better than RGB.

The threshold value of either RGB or HSV can simply be the average value. However, if the whole point cloud has been used in averaging, the thresholding would not be very efficient since the lighting of the whole leaf can be different in some part. Thus, the limiting the number of points to be only around the interested part of the leaf. To do so, the proposed system took the coordinates of 2 points of leaf edges, one from left and one from right, in the part we want to measure and make the linear equation L out of them. The next step is to assign the number of total points n which we want to limit to and find n point that are closest to the L that has been calculated earlier. The thresholding will be performed only on these n points for both RGB and HSV and, obliviously, the results from these two are not necessary to be the same. If RGB and HSV thresholding are both state the same single point is the leaf vein, we label the point as the leaf vein. On the other hand, If the two distinguish points are selected by each thresholding, we will not label any of them as the leaf vein. But the problem is sometimes these two distinguish point from RGB thresholding p_1 and from HSV thresholding p_2 are very close to each other, it would not be fair to label them as a leaf surface. This can be solved by measuring the average distance from p_1 to the 10 closest point and compare it to the distance between

p_1 to p_2. If the distance between p_1 to p_2 is shorter, we can assure that these two points are close to each other and then can be safely labelled as the leaf vein.
After the thresholding, the vein point can be obtained and now the distant between each leaf vein can be measured. But as there could be several points in a single leaf vein, we decided to use the centroid of each individual vein in the measurement. At this stage, the measured width will not be in the metric unit and thus cannot be understandable the true dimension. To tackle this problem, the system measures the width of one block of the scale reference, which each one has a length of one centimetre, and compare it with the obtained veins distance to convert it into the centimetre unit. By this approach, the system can finally generate the true dimension of a leaf in centimetre that can be understandable and validated with the real plant.

3. Experimental result

In this paper, the experiment for validating our approach was conducted on a fake rice tree model made of paper, shown in figure 2(a). The model has been created with metal wires for stems and green papers for leaves. The wires were wrapped with the green paper tape and the paper leaves were wrote with green ink for making the leaf veins.

(a)

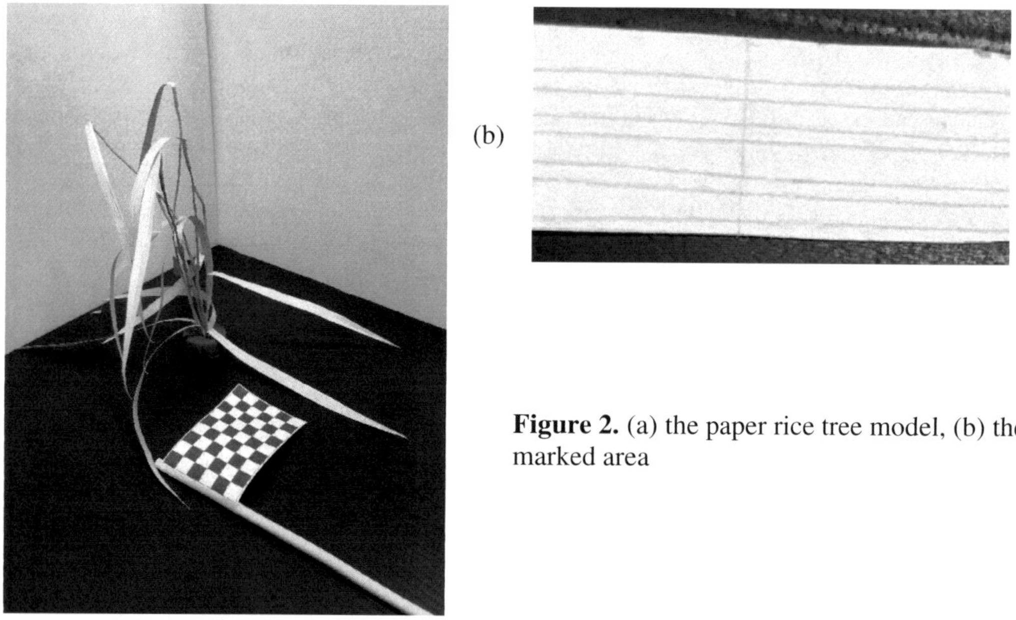

(b)

Figure 2. (a) the paper rice tree model, (b) the marked area

There are several reasons why the authors conducted an experiment on paper model. First, the verifying the result would be a lot easier than doing with the real rice plant since it is not growable, which mean we do not have to concern of the growing of the leaf in the meantime the experiment is being conducted and validating afterward could be inaccurate. Second, the setting up of an experiment is much easier than the real rice as it will not break apart easily or get affected by external environment.
In this experiment, the authors marked a faint line across the specific leaf where the measurement is going to be made on with a pencil, shown is figure 2(b). The leaf was recorded along with the scale reference in a form of chessboard pattern flag with an iPhone 7 camera in 4k 30fps for 15 seconds. From the video, we extract it into 159 pictures, each size 3840 x 2160 pixels, in a form of video frame and use them in the 3D reconstruction with COLMAP. The reconstructed point cloud contains 308,861 points as shown in figure 3.

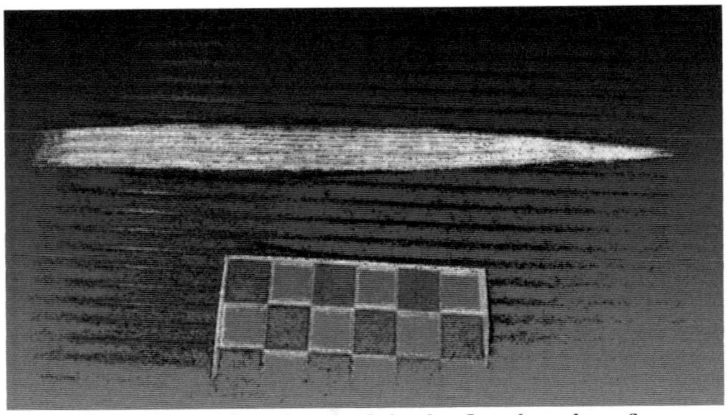

Figure 3. The point cloud of the leaf and scale reference

After obtaining the point cloud, two points were selected whereas located in the left and right leaf edge on the marked area and created the linear equation from these points. Then, the 1,500 nearest points to the line were selected and used in thresholding. The thresholding results are displayed as the red points in figure 4. The measured distances are then calculated and compared with the results from Easy Leaf Area (ELA) [2] which has a similar idea but executed in 2D top view image.

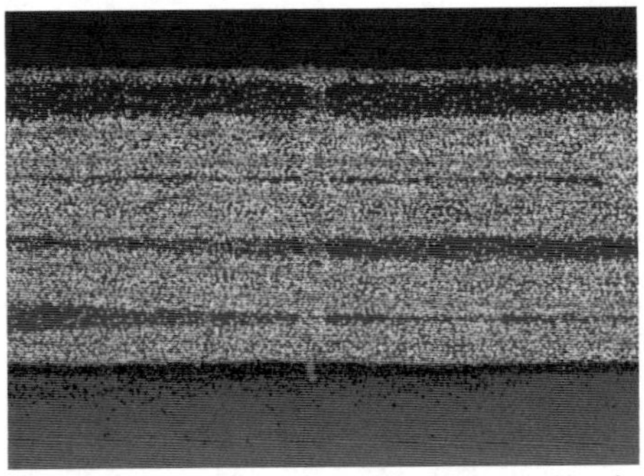

Figure 4. The thresholding result (red points)

The pictures we used for measuring with ELA contains three pictures: the direct top view picture, a picture taken with a tilted camera angle, and a picture where the leaf is not on the same level as the scale reference. These pictures are shown in Figure 5.

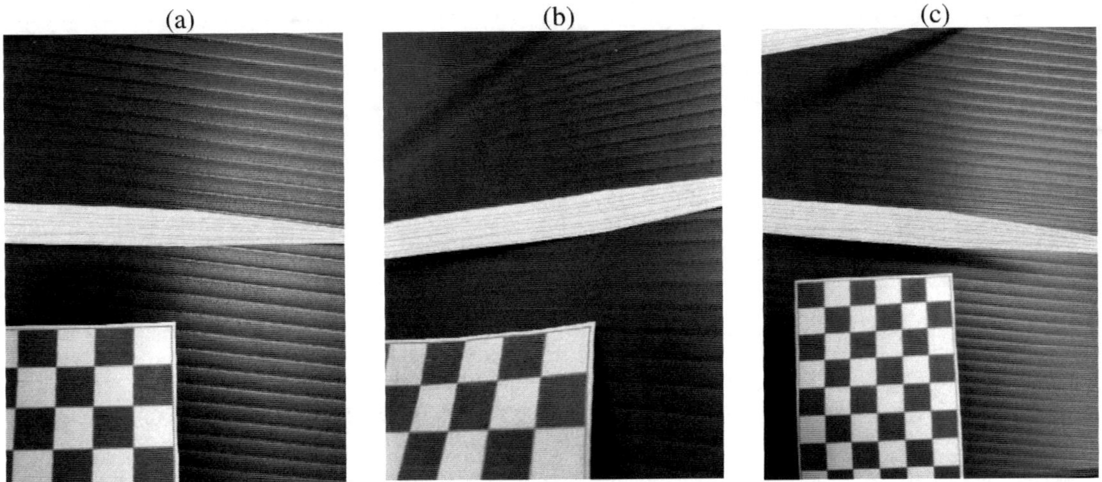

Figure 5. The picture of a leaf taken with different condition: (a) direct top view, (b) tilted camera angle, (c) the leaf level is higher than the scale reference

The leaf model in this experiment contains 9 veins as shown in figure 6. The measuring result between each vein of each picture comparing with the measuring in point cloud is shown in table 1.

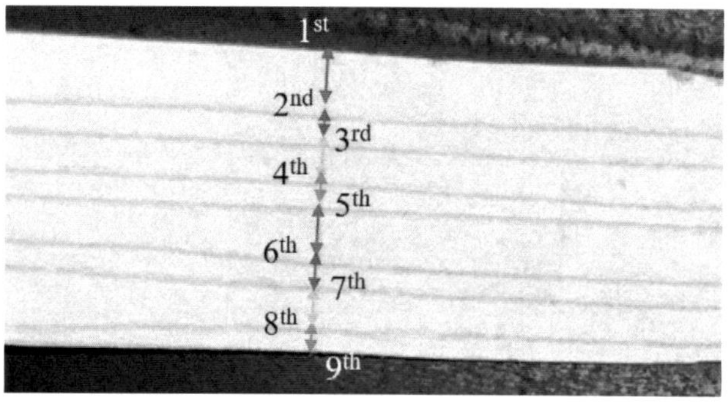

Figure 6. the number of each vein

Table 1. The measuring result between each vein for each experiment in millimetres

Distance between veins (mm)	1st to 2nd	2nd to 3rd	3rd to 4th	4th to 5th	5th to 6th	6th to 7th	7th to 8th	8th to 9th
Ground truth	2.05	1.03	1.25	0.86	1.83	0.86	1.45	0.72
2D: Top view	1.96	0.89	1.19	0.72	1.66	0.78	1.27	0.67
2D: Tilt angle	1.59	0.68	0.93	0.66	1.30	0.68	1.05	0.54
2D: Different level	3.10	1.37	1.82	1.20	2.61	1.11	2.04	0.97
3D: Point cloud	2.17	1.04	1.19	0.90	1.82	0.78	1.45	0.81

4. Discussion

The accuracy of Easy Leaf Area is highly depending on the camera angle and the level between the scale reference and the associated leaf as evidenced in the experiments where the camera angle is tilted and when the scale reference is place lower than the leaf. On the other hand, 3D space allows the independency of camera angle and makes the accuracy of our approach the highest among all experiments. The root mean square error (RMSE) for each experiment is shown in table 2.

Table 2. Root mean square error for each experiment

Experiment	2D: Top view	2D: Tilt angle	2D: Different level	3D: Point cloud
RMSE	0.0115	0.0326	0.0518	0.0051

For the 2D measured distance between leaf veins, direct top view image is the most accurate among 2D image. The curvature of the leaf makes the result slightly incorrect. For the tilt angle image,
The tilt angle causes the reference pattern to be more toward the camera while the leaf is further and affect the conversion of unit. In different level image, because of the different level between the leaf and the pattern makes the leaf look bigger, which leads to the larger measurement result than a ground truth.
For 3D point cloud measurement, there are slight errors at the distance between veins which may caused by the thresholding result. As can be seen in figure 4, all the red points are not perfectly aligned in the form of linear line while the measuring distance between veins at the linear line is the way the ground truth is created. Despite the error, the results indicate that this approach is still more accurate than using only 2D images.
Although the cell phone camera, specifically iPhone 7 with 12-megapixel resolution camera, is used in this study, there is a possibility that the result could get more accurate when using the professional camera instead of a cell phone camera since the image resolution is higher which can lead to finer surfaces of the reconstructed point cloud. However, as the number of people with smart phone is much

higher than the people who have professional camera, the authors decided to use the cell phone camera as it makes the system more compact and can be easily used by any person.

5. Conclusion

The system uses 3D reconstruction techniques to measure the distance between leaf veins by taking a set of 2D images of a rice leaf with SfM and MVS to generate 3D point cloud. From the point cloud, we extract venation with RGB and HSV thresholding, and lastly, converting the measured unit into millimetres with the help of a scale reference. Our system successfully generates the 3D point cloud and estimate the width between each vein with higher accuracy than the result from using 2D images. As the method for the leaf veins classification in point clouds based on a thresholding result introduced in this paper still has a room for improvement in term of the accuracy, the authors plan to propose a better way to selecting the leaf veins without heavily depending on the colour information in future.

6. References

[1] Maloof J N, Nozue K, Mumbach M R and Palmer C M 2013 LeafJ: an ImageJ plugin for semi-automated leaf shape measurement. *JoVE (Journal of Visualized Experiments)* (71), e50028.

[2] Easlon H M and Bloom A J 2014. Easy Leaf Area: Automated digital image analysis for rapid and accurate measurement of leaf area. *Applications in plant sciences* 2(7), apps.1400033. https://doi.org/10.3732/apps.1400033

[3] Vázquez-Arellano M, Reiser D, Paraforos D S, Garrido-Izard M, Burce M E C and Griepentrog H W 2018 3-D reconstruction of maize plants using a time-of-flight camera. *Computers and Electronics in Agriculture* 145 pp 235-247.

[4] Weiss U and Biber P 2011 Plant detection and mapping for agricultural robots using a 3D LIDAR sensor. *Robotics and autonomous systems*, 59(5), pp 265-273.

[5] Seidel D, Beyer F, Hertel D, Fleck S and Leuschner C 2011 3D-laser scanning: A non-destructive method for studying above-ground biomass and growth of juvenile trees. *Agricultural and Forest Meteorology,* 151(10), pp 1305-1311.

[6] Ullman S 1979 The interpretation of structure from motion. *Proceedings of the Royal Society of London. Series B. Biological Sciences*, 203(1153), pp 405-426.

[7] Bailer C, Finckh M and Lensch H P A 2012 Scale Robust Multi View Stereo. *Computer Vision – ECCV 2012. Lecture Notes in Computer Science*, vol 7574

[8] Schönberger J L, Zheng E, Frahm J M and Pollefeys M 2016 Pixelwise view selection for unstructured multi-view stereo. In: *European Conference on Computer Vision*. Springer, Cham, pp 501-518.

[9] Schönberger J L and Frahm J M 2016 Structure-from-motion revisited. In *Proceedings of the IEEE Conference on Computer Vision and Pattern Recognition* pp. 4104-4113.

[10] Furukawa Y, Curless B, Seitz S M and Szeliski R 2010Towards internet-scale multi-view stereo. In: *2010 IEEE computer society conference on computer vision and pattern recognition.* pp 1434-1441.

Acknowledgments
We would like to thank you Hiroki Yano for helping us beautifully create the rice leaf model and assisting in numerous experiments.

Methods for verification of groundwater flow simulators (case study: Jakarta groundwater basin in the urban villages of Jatiluhur and Jatirangga, Bekasi City)

T Rahmawati[1], G L Putri[2], D R Marthanty[1], F Zulkarnain[3], C R Priadi[1,2]*

1 Civil and Environmental Engineering Department Faculty of Engineering, Universitas Indonesia, Depok, 16424, Indonesia
2 Center for Sustainable Infrastructure Development, Faculty of Engineering, Universitas Indonesia, Depok, 16424, Indonesia
3 Geography Department, Faculty of Mathematics and Natural Science, Universitas Indonesia, Depok, 16424, Indonesia
E-mail: crpriadi@ui.ac.id

Abstract. Limited access to clean water leads to residents extracting groundwater through dug wells or boreholes, which in turn results in groundwater becoming contaminated. Groundwater flow simulation of the Jakarta basin, which has been previously modeled with the *Groundwater Modeling System 10.3,* can be used as a basis for simulating groundwater flow pollution. Before simulating the groundwater flow pollution, the groundwater flow simulator needs to be verified to examine whether simulators and conditions in the field correspond to the water table level. This study demonstrated the methodology used for verification of the simulator in research. Verification was done using the Chi-square method. The results revealed that calculated Chi-square values were larger than Chi-square table values for the two urban villages of Jatiluhur and Jatirangga. The value of calculated Chi-square value for Jatiluhur was 17.3, whereas it was 7.81 in the Chi-square table. In Jatirangga, the calculated Chi-square value was 37.23, whereas it was 14.1 in the Chi-square. Results indicated that the simulation was not in accordance with the actual water table level. This discordance can occur since it necessary to update the data on groundwater flow simulation results with bore log data or to verify the hydrodynamic conditions of the study area.

1. Introduction

Utilization of groundwater for daily needs is a factor affecting groundwater balance. Changes in groundwater level due to uncontrolled groundwater extraction can result to a reduction in the age of deep groundwater (in confined areas), leading to surface water entering deep groundwater and contaminating deep groundwater [1]. In addition, residents can also face difficulties in accessing clean water from groundwater because the groundwater level of the well decreases significantly, whereas the supply of clean piped water in the area might still be limited.

In groundwater basins, fundamental components include recharge and discharge areas. To study the behavior of real systems in the field, it is necessary to simulate the process of implementing the model into a computer program and execute it in a manner that resembles the system in the field [2]. Previously, a groundwater management simulation was conducted in the Jakarta groundwater basin [1]. The aim was to simulate the groundwater hydrodynamic condition of the Jakarta groundwater basin in order to conduct various policy simulations related to groundwater contour. The simulator refers to the Jakarta groundwater basin, which includes Bekasi City in the urban villages of Jatiluhur and Jatirangga. The simulator was made using the Groundwater Modeling System (GMS) software 10.3. Modules used included 2D Grid, 3D Grid, and Borehole, whereas the model used was MODFLOW. The input parameters of the simulator manufacture consisted of horizontal and vertical boundary conditions, soil contour and ground water level, initial conditions, general head values, specific head values, aquifer hydraulic parameters, river boundaries, lakes, recharge wells, and groundwater extraction. The data for these parameters were obtained from secondary data, which was then used to simulate the hydrodynamic process that occurs in the Jakarta groundwater basin.

Jatiluhur and Jatirangga, two urban villages in Bekasi City, are included in the Jakarta groundwater basin. The Jakarta groundwater basin is a basin that is bounded by a free groundwater level at the top and a tertiary-aged rock (impermeable) at the bottom [3]. The northern part of the Jakarta groundwater basin is bounded by the Java Sea with a constant head boundary value of 0; in the western part, there is the Cisadane River, which is included in the no flow boundary; furthermore, the eastern part is bounded by the Cikeas River and the Bekasi River, and the southern part is bounded by rocks that arise in the Bulak Kulon area encompassing Depok City and which is considered as the groundwater flow limit because groundwater flow originating from the south of Depok is relatively small at 1 m^3 per year [1].

Model verification and simulation are required to assess modeling accuracy and uncertainty [4] and to measure the suitability of the simulation results in terms of whether they approach field conditions or not. The more the simulation results correspond with field conditions, the better the simulator is in describing field conditions. Generally, modeling verification is done using statistical comparison methods. One method includes the Chi-square method. A simulator, such as the one utilized by Apriatresnayanto in 2018 [1] is necessary to see the actual conditions on the ground. The purpose for the verification of this groundwater flow simulator is to analyze the results of the simulator and calibrate the model with newer data, including more pumping test data to obtain better aquifer parameter results. The purpose of this study is to describe the method used to verify simulation results and field conditions using the Chi-square test.

2. Materials and methods

2.1. Groundwater level measurement

Groundwater has certain characteristics in terms of its movement patterns, seepage, and other features. Groundwater flows with a certain movement pattern and velocity, and differences in soil properties between permeable media and other permeable media [5]. Groundwater level refers to the surface of a groundwater body that is pressurized by atmospheric air [6]. Based on SNI 7749: 2012 procedures for determining groundwater levels in boreholes or monitoring wells, there are three measurement methods: a ballast gauge, a sound measuring instrument, and an electric flame meter. However, measuring instruments with sound marks can only be utilized in boreholes or wells that have shallow ground water levels. Groundwater level gauges have both advantages and disadvantages. Measurement with weights produces the highest accuracy; however, the measuring tape used must clearly reflect the difference between wet and dry tape.

2.2. Research sites

The study sites included the Jakarta groundwater basin, specifically the Jatiluhur and Jatirangga urban villages in Bekasi City, West Java (hereinafter referred to as Jatiluhur and Jatirangga). The selection of research sites was based on several considerations and conditions, including the level of poverty seen at the sub-district level and at the urban village level, population density at the sub-district level, hydrogeological conditions, such as the surface conditions below which are still included in the Jakarta groundwater basin, aquifer productivity, and distance of pollutant sources to groundwater sources. Four points in the Jatiluhur area and eight points in the Jatirangga area were included. Figure 1 shown the points on Bekasi City. The orange area is Jatiluhur, and purple area is Jatirangga.

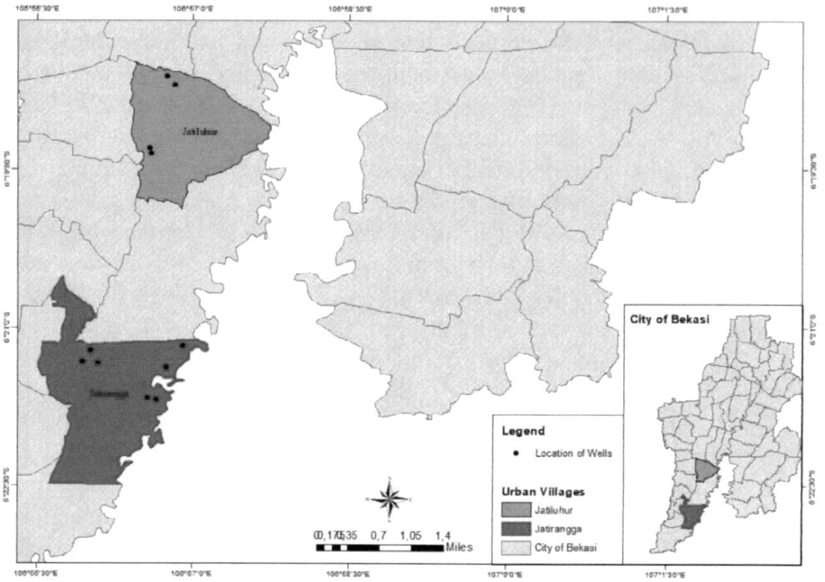

Figure 1. Research Location

2.3. *Regional condition of research area*

The bottom of the aquifer system of the Jakarta groundwater basin is formed by opaque Miocene sediments, which are also outcrops in the southern boundary of this basin. The Jakarta groundwater basin consists of sea Pliocene and Quaternary sand and delta sediments up to 300 m. Quaternary sediments are divided into three aquifer systems based on hydraulic characteristics and depth, namely, phreatic aquifer systems (0 to -40 m), upper confined aquifer systems (-40 m to -140 m), and bottom depressed aquifer systems. The average horizontal hydraulic conductivity value is approximately 1×10^{-3} cm.s^{-1}, and the hydraulic conductivity value is roughly 1×10^{-5} cm.s^{-1} [5]. In Figure 2, the yellow area represents the Jakarta groundwater basin.

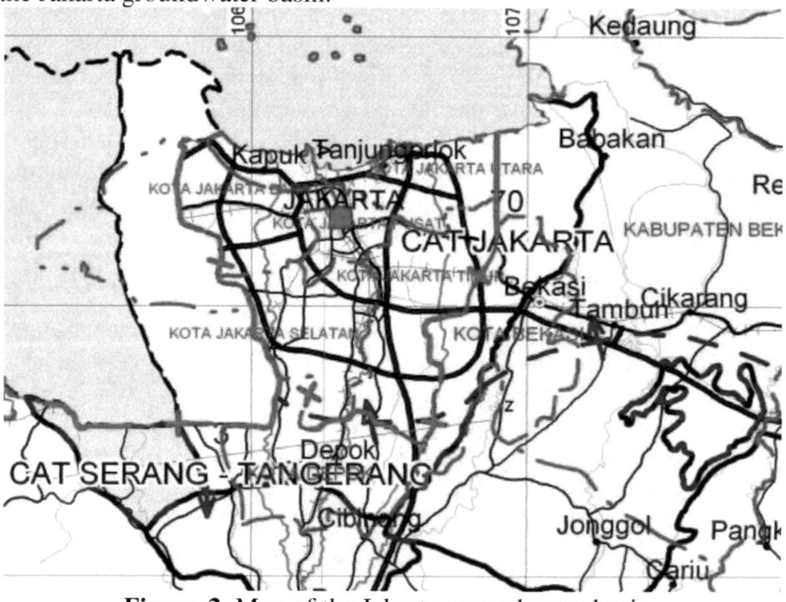

Figure 2. Map of the Jakarta groundwater basin.

2.4. Chi-square test

Verification is a necessary step for model acceptance [6]. One verification method includes the Chi-square test. The Chi-square test is a non-parametric statistical test. Hypothesis testing for differences in more than two proportions of a population cannot use T or F distribution; however, it can use the Chi-square distribution. The Chi-square test belongs to the class of statistical hypothesis-testing mechanisms and is commonly used to determine whether the assumed mean and covariance matrices match the actual settings or not [7].

2.5. Data

At the research locations in Jatiluhur and Jatirangga, Bekasi City, it was found that there were 12 dug wells that could be measured in depth using fiberglass measuring tapes, and their coordinates were recorded. Of the 12 dug wells, four of them were located in Jatiluhur, and eight of them were located in Jatirangga. The depth of the wells in the field varied from 9 to 29 m. Based on the depth of the well, this value was compiled with GPS-recorded data relating to altitude in the field. The results of measurements in the field are shown in Table 1. The HH ID column shows the measurement point name for the depth of the dug wells at the study site. The depth of the groundwater level was obtained from the altitude (Z) value minus the measurement results of the well depth.

Table 1. Coordinates and depth of groundwater level

Urban Villages	Sample Code	Latitude (East, X)	Longitude (North, Y)	Altitude in Field (Z)	Depth of Well (meter)	Water Table Level (meter)
Jatiluhur	JL1	715414	9301945	79.7	9	70.7
	JL2	715282	9302101	57.5	10	47.5
	JL3	715011	9300747	78.5	14	64.5
	JL4	714980	9300835	85.0	12	73.0
Jatirangga	JR1	713749	9297110	88.3	12	76.3
	JR2	713759	9297104	60.7	18	42.7
	JR3	714028	9297075	70.3	7.5	62.9
	JR4	713896	9297305	72.5	12.3	60.5
	JR5	715080	9296424	72.2	17	55.2
	JR6	714921	9296452	84.6	29	55.6
	JR7	715251	9296994	82.0	12	70.0
	JR8	715548	9297369	80.2	11.2	69.0
Mean			Jatiluhur	75.2	11.3	63.9
			Jatirangga	76.4	14.8	61.5
Standard Deviation			Jatiluhur	12.1	2.2	11.5
			Jatirangga	9.0	6.6	10.5

The next dataset used included head values from the results of Jakarta groundwater basin flow simulator that were collected previously by Apriatresnayanto in 2018 with *Groundwater Modelling System 10.3* [1]. The results of the simulators used are from simulations with groundwater extraction scenarios using residents' dug wells or borehole. Groundwater retrieval by residents mostly occurs in buffer areas, such as the cities of Depok, Bekasi, and Tangerang, which can cause the pattern of groundwater contour to change when compared to systems without groundwater uptake. Intake of groundwater from the wells of these inhabitants affects basins in several places. The simulation results are depicted in Figure 3. The simulation of groundwater basin flow can be a basis for building the simulation for contaminant flow in

IOP Conf. Series: Earth and Environmental Science **633** (2021) 012014 doi:10.1088/1755-1315/633/1/012014

groundwater. Tracer for contaminant flow in the simulation refers to the concentration of total dissolved solids (TDS) in residential areas, which describe pollutants generated from population sanitation facilities, such as septic tanks and ponds connected to toilets for fecal containment.

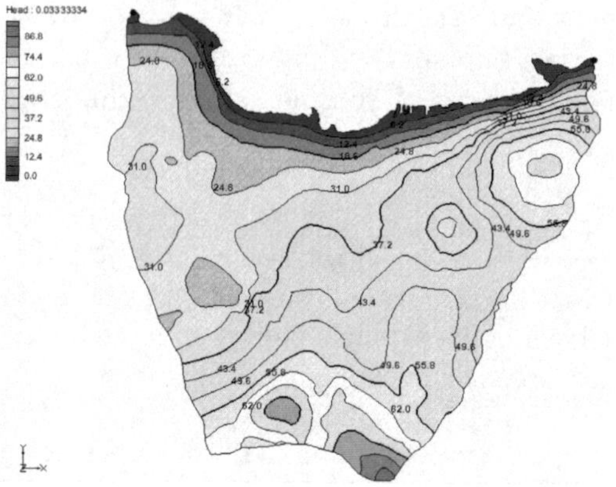

Figure 3. Simulation results of groundwater flow using groundwater near a community well (shallow well) [1].

3. Results

The verification test can be conducted by comparing the results head of the groundwater flow simulator of the Jakarta groundwater basin and the results of the water table level measurements at the several points included in the groundwater basin. Head is the water table level of the groundwater flow simulation. Twelve dug wells were measured at groundwater level. For the locations of the 12 dug wells, the head values were determined from the results of the simulation of ground flow, which had been previously simulated with the help of GMS 10.3 (MODFLOW) software [1]. Based on the results of water table level in field measurements, significant differences compared to the results of the simulator were observed. Head of simulation values tended to be stable at values ranging from 45.0 to 49.6 m, whereas the depth of the groundwater level ranged from 42.7 to 76.3 m. The differences between head of simulation and depth of the groundwater level values are depicted in Table 2.

Table 2. *Head* from simulation and groundwater level depth

Urban Villages	Sample Code	Water Table Level (meter)	*Head of Simulation*
Jatiluhur	JL1	70.7	49.6
	JL2	47.5	49.6
	JL3	64.5	49.6
	JL4	73.0	49.6
Jatirangga	JR1	76.3	45.0
	JR2	42.7	45.0
	JR3	62.9	47.0
	JR4	60.5	45.0
	JR5	55.2	45.0
	JR6	55.6	45.0
	JR7	70.0	49.6
	JR8	69.0	48.0
Mean	Jatiluhur	63.9	49.6

	Jatirangga	61.5	46.2
Standard Deviation	Jatiluhur	11.5	0.0
	Jatirangga	10.5	1.8

The altitude of the field, the groundwater level, and the simulated groundwater level demonstrated trends that were mutually sustainable with each other. Figures 4 and Figure 5 shown the comparison of these parameters in Jatiluhur and Jatirangga.

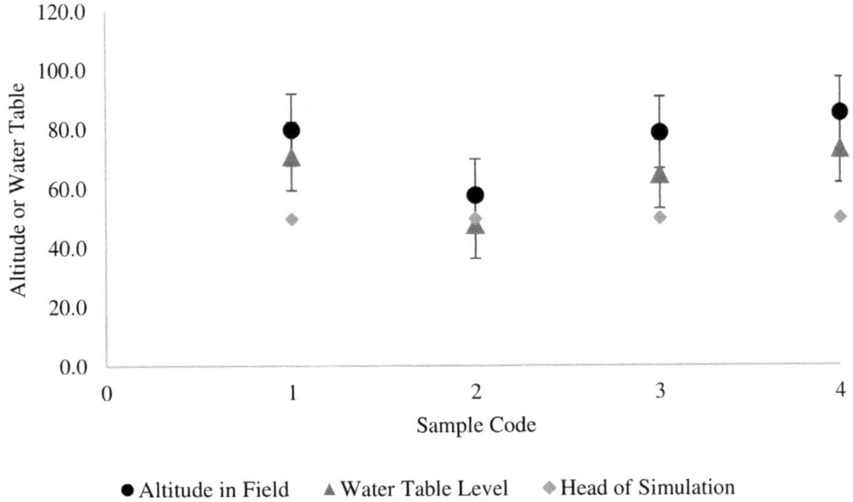

Figure 4. Comparison of Altitude in Field, Water Table Level, and Head of Simulation in Jatiluhur Urban-Village

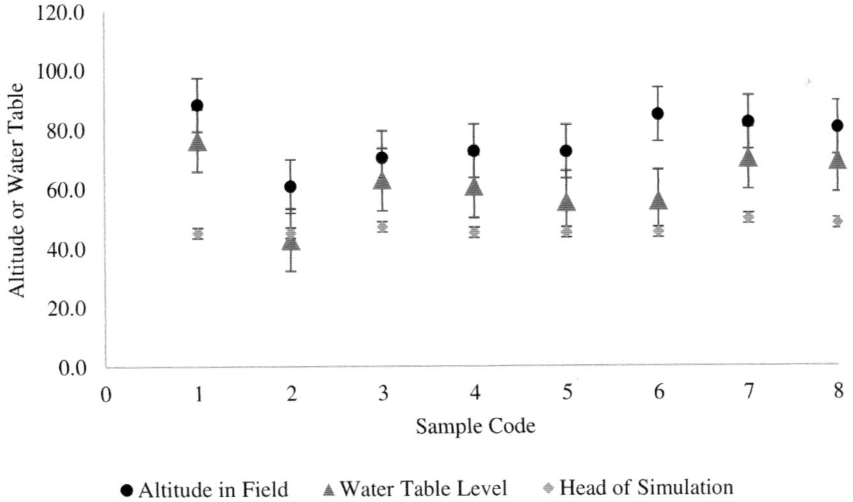

Figure 5. Comparison of Altitude in Field, Water Table Level, and Head of Simulation in Jatirangga Urban-Village

The difference in the head of the simulation results compared with the depth of the groundwater level in the field was then subjected to a verification test using the Chi-square method. Before doing the calculations, it was first necessary to make a hypothesis. The hypothesis consists of H_0 and H_1. H_0 states that the simulation results data are in accordance with the conditions in the

IOP Conf. Series: Earth and Environmental Science **633** (2021) 012014 doi:10.1088/1755-1315/633/1/012014

field, whereas H_1 states that the simulation results is not in accordance with the conditions on the ground. There are two rules used in this method. The first rule states that if the Chi-square count results are smaller than those of the Chi-square table, H_0 is accepted, and H_1 rejected. The second rule states that if the Chi-square count is greater than the Chi-square table, then H_0 is rejected, and H_1 is accepted.

These hypothesis was used as a basis for decision making of verification result. The depth of the groundwater level was compared with the head value resulting from groundwater flow simulation. After that, a verification test was conducted using the Chi-square method and compared with the Chi-square values contained in the table. Chi-square calculation was done according to Equation 1:

$$\chi^2 = \sum \frac{(O_i - E_i)^2}{E_i} \quad (1)$$

Where the O_i is an observation data value (simulation data - i), and E_i is an expected data value (theoretical data or the *i*-th real system).

After obtaining the Chi-square values from the calculation, it was then compared with the Chi-square values from the table according to the degree of freedom and significance level. The degree of freedom was obtained from the number of samples minus 1 (n-1). In this study, verification tests were conducted on 12 dug wells. The degree of freedom was 11. The level of significance, also known as α, is a measure of how much confidence will be taken. The value of α used was 0.05, and the confidence level of the decision taken was 95%.

The calculated Chi-square values and Chi-square table values are depicted in Table 3. Results revealed that calculated Chi-square values were larger than Chi-square table values for both Jatiluhur and Jatirangga. The calculated Chi-square value for Jatiluhur was 17.3, whereas it was 7.81 for the Chi-square table. In Jatirangga, the calculated Chi-square value was 37.23, whereas for the Chi-square table, it was 14.1. Such results mean that H_0 is rejected, and H_1 is accepted; in short, the simulation was not in accordance with the water table level. This can occur because it is necessary to update the data on groundwater flow simulation results with bore log data or to verify the hydrodynamic conditions of the study area.

Table 3. Value of Chi-square Calculated and Chi-square Table

Sample Code	Chi-square Calculated		Chi-square Table
JL1	6,29	17,30	7,81
JL2	0,09		
JL3	3,42		
JL4	7,50		
JR1	12,83	37,23	14,1
JR2	0,13		
JR3	4,04		
JR4	3,99		
JR5	1,89		
JR6	2,03		
JR7	5,93		
JR8	6,40		

4. Conclusion

Verification test results of the simulation and field conditions can be conducted using the Chi-square method through manual calculation. Before calculating the value of Chi-square, it is necessary to formulate a hypothesis in advance regarding the relationship between the calculated Chi-square values and Chi-square table values. The results of the current study indicate that calculated Chi-square values were larger than Chi-square table values for both Jatiluhur and Jatirangga. This means that H_0 is rejected and that H_1 is accepted, such that the simulation is not accordance with the water table level. To ensure that simulation results are in accordance with field conditions, the calculated Chi-square value must be smaller than the Chi-square table value. There are a few reasons why the simulation may be discordant with the actual water table level. This can occur when data on groundwater flow simulation results is not updated with bore log data or when the verification of hydrodynamic conditions of the study area and simulators has not been conducted.

5. References

[1] Apriatresnayanto R 2019 Jakarta Groundwater Basin Recharge - Discharge Boundary Area Map: A Preliminary Study. *IOP Conf. Series: Materials Science and Engineering* 690

[2] Sridadi B 2009 *Modelling and Simulation System*, (Jakarta: Informatika).

[3] Mukhtar O, Pranantya P A and Hadian S D 2019 Groundwater Management in DKI Jakarta Groundwater Basin. *3rd National Seminar of Geological Engineering Universitas Padjadjaran*

[4] McFarland J and Mahadevan 2008 S Multivariate significance testing and model calibration under uncertainty *Computer Methods in Applied Mechanics and Engineering* 197 p 2467-2479.

[5] Simaremare S 2015 One-Dimensional Groundwater Flow Analysis *Electronic Journal of Civil and Environmental Engineering, Unsri* 783-794.

[6] National Standardization Agency of Indonesia. 2012 Indonesian National Standard Number 7749 - Procedure for Determination of Groundwater Surface at the Borehole or Monitor Well (Jakarta: BSN Indonesia)

[7] Mayer D and Butler D 1993 Statistical Validation *Ecological Modelling* 21.

Acknowledgements
This research was funded by the Water for Women Fund from the Australian Government Department of Foreign Affairs and Trade (number WRA1:1004) and also by PTUPT from the Ministry of Research, Technology and Higher Education of Indonesia (number: ADD-NKB-2943/UN2.RST/ HKP.05.00/2020). We thank Juliet Willetts, Tim Foster, and Fransziska Genter from University of Technology Sydney and Angela Harris from North Carolina State University for their extraordinary discussion, knowledge sharing, and collaboration in this research.

Health risk analysis of nitrite, nitrate, and heavy metal pollution in groundwater near landfill area: A case study of the Sumur Batu village in Bekasi, Indonesia

WH Pratiwi[1], GL Putri[2], MA Pratama[1], F Zulkarnain[3], and CR Priadi[1,2] *

1 Civil and Environmental Engineering Department, Faculty of Engineering, Universitas Indonesia, Kampus UI Depok, 16424, Indonesia
2 Center for Sustainable Infrastructure Development, Universitas Indonesia, Kampus UI Depok, 16424, Indonesia
3 Geography Department, Faculty of Mathematics and Natural Science, Universitas Indonesia, Kampus UI Depok, 16424, Indonesia
E-mail: crpriadi@ui.ac.id

Abstract. Previous studies state that improper landfill or waste management is one of the main causes of nitrite, nitrate, and heavy metal contamination in groundwater. In Bekasi, Indonesia, 70% of the population depends on groundwater for daily use, including the residents of Sumur Batu. The village of Sumur Batu is located adjacent to two landfills, namely, the Bantar Gebang landfill, which is the largest landfill in Indonesia, and the Sumur Batu landfill. This research analyzed the health risk of drinking groundwater that may have been contaminated by nitrite, nitrate, and heavy metals. Data processing was performed separately for different age groups: infants, children, and adults. The results showed that there is a risk of NO_2 exposure in all age groups when groundwater is used as a source of drinking water in Sumur Batu. Because the region is directly bordered by a landfill area and the community is highly dependent on groundwater, as long as access to water services is not guaranteed, the quality of groundwater in this area needs to be monitored.

1. Introduction

Water is a crucial basic human need, and one-third of the world's population depends on groundwater to meet their daily needs [1]. Previous studies state that improper landfill or waste management is one of the main causes of nitrite, nitrate, and heavy metal contamination in groundwater [2][3]. The city of Bekasi has relatively high population growth, which is causing some people to face difficulties in accessing an adequate supply of safe water. Only 26.8% of Bekasi City's water needs are met by the government [4]. Sumur Batu is one of the urban villages in Bekasi City that is very dependent on groundwater because it hasn't yet obtained guaranteed water access from the government. This village is directly adjacent to two landfills, namely, the Bantar Gebang landfill and Sumur Batu landfill.

The Bantar Gebang landfill is the largest landfill in Indonesia, receiving up to 7,000 tons of garbage from the city of Jakarta per day. This 110.3 Ha landfill has been operating since 1989 with a sanitary landfill treatment system [5]. However, the waste at Bantar Gebang Landfill is not professionally handled [6], and its leakage has caused water pollution and air pollution [7]. The Sumur Batu landfill, which has an area of 21 ha, is located in Sumur Batu village and receives 700–900 tons of garbage per day from 12 sub-districts of Bekasi. The waste management system at the Sumur Batu landfill involves open dumping, which has a great potential to cause various types of pollution.

Although many households in Sumur Batu are dependent on groundwater, its quality is declining. The groundwater in the area around the Bantar Gebang landfill is likely to have been polluted [8]. This is also evidenced by several previous studies, including research which states that the average contents of Cr, Zn, Fe, Mn, and Hg in all wells sampled around the Bantar Gebang sub-district of Bekasi are 2.12 mg/l, 7.34 mg/l, 0.86 mg/l, and 13.31 mg/l, respectively. Those values are above the quality threshold for class 1 water (i.e., clean water used for drinking) defined by Government Regulation No. 82, enacted in 2001[9]. Using the STORET system from the US Environmental Protection Agency (US EPA), Bethy (2011) shows that the well in the Bantar Gebang sub-district has poor water quality and is affected by the Bantar Gebang landfill [10].

Leachate produced by solid waste contains many organic and inorganic compounds. Based on Irhamni's (2017) research, the leachate from landfill can contain heavy metals, such as mercury (Hg), copper (Cu), iron (Fe), cadmium (Cd), cobalt (Co), zinc (Zn), nickel (Ni), chromium (Cr), and lead (Pb). Additionally, the decomposition of organic matter by bacteria will lead to high nitrate and nitrite content in the waters [11]. The entry of leachate into groundwater and its distribution is expected to occur through the process of osmosis, capillarity mechanisms and electrokinetic processes [12].

In southwest Nigeria, it was reported that the presence of heavy metals in groundwater can significantly increase the health risks of cancer and non-cancer [2]. While, a high content of nitrates in drinking water is associated with methemoglobinemia, which is a great risk for infants under 6 months [13]. Additionally, a study in India shows that the location of municipal waste disposal broadly contaminated Chandigarh's groundwater. The average risk quotient (RQ) for As, Co, Cd, Cu, Ni, Pb, and Zn (but not for Cr) is >1. Therefore, consumption of Chandigarh groundwater has the potential to cause non-carcinogenic health impacts on living things [3].

2. Materials and method

2.1 Study area

The study was located in Sumur Batu, Bantar Gebang, Bekasi. Sumur Batu has an area of 5.7 km^2 and a population of 17,002 people. Residents mainly obtain groundwater from a borehole. The study location was selected used a purposive sampling method based on several criteria: high use of groundwater as a source of drinking water, a lack of piped water connection, high poverty rate, high population density, and proximity to landfills. A total of 24 groundwater samples were taken based on random sampling in Sumur Batu, a map of the sample distribution at the study locations is shown in Figure 1.

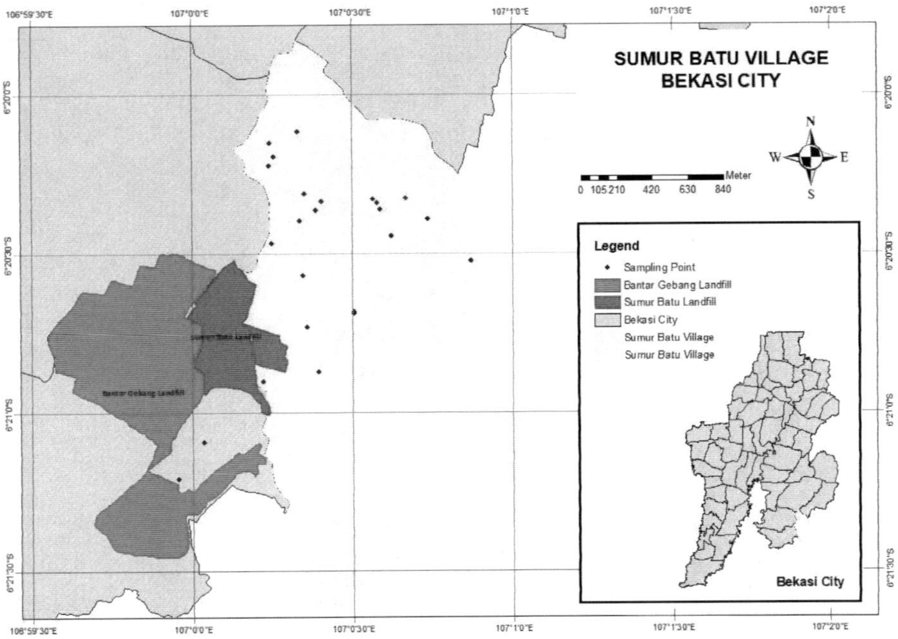

Figure 1. Study site area (Sumur Batu) and sampling point distribution

2.2 Sampling and analysis

Water sampling from wells in households was carried out at 24 sample points in the rainy season, from 2–14 March 2020. Groundwater samples were collected from boreholes with a depth of 10–15 m below the ground surface. The high-density polyethylene (HDPE) bottles, which were used as a water sample container, were washed three times with the sample before the sample was collected. Each sample was put into two bottles, one bottle for the nitrite and nitrate test and another for the heavy metal test. The pH and total dissolved solids (TDS) were measured in the field using a multiparameter (HANNA-HI9828) that was calibrated beforehand. Then, samples were preserved based on SNI 06-6989.16-2004. Each nitrite-nitrate sample was acidified in situ with 2 M H_2SO_4 until a pH of <2 was achieved. Heavy metal samples were also acidified with 2 M HNO_3 until a pH of <2 was achieved in order to dissolve all the heavy metals and prevent crystallization or precipitation and adsorption of trace metals to the container surfaces before analysis. All water samples were kept at 4°C in a sealed cooler and then transported to the laboratory for analysis of the concentrations of nitrite and nitrate by a spectrophotometer and the concentrations of Cr, Zn, Fe, Mn, Hg, and Cd by inductively coupled plasma mass spectrometry (ICP-MS) and atomic absorption spectrophotometry (AAS).

2.3 Questionnaires

This study is focused on households who consume shallow groundwater in Sumur Batu. Based on the Slovin method of considering the population number, the minimum number of samples (in this case, respondents) who can represent Sumur Batu is 391 people. In this study, 143 households (597 residents) were randomly chosen. This amount is consideread to represent the population of the Sumur Batu. Face-to-face interviews using a questionnaire were carried out with those households in the same period as water sampling, 2–14 March 2020. The age range was 0–90 years. The questionnaire asked for general information about the participants and their groundwater consumption behavior, such as the intake rate, frequency, and quantity of consumption.

2.4 Health risk assessment

Health risk assessment is a method for calculating the carcinogenic and non-carcinogenic risks to human beings after chemical exposure. The ingestion of contaminated groundwater can adversely affect human health through a variety of methods of exposure, including direct ingestion, dermal contact, and washing [14]. However, this study only considers the effects of groundwater consumption on human health as an exposure pathway. The absorption of potential toxins through drinking water can be expressed by chronic daily intake (CDI), which is calculated as follows [14]:

$$CDI = C \times IR \times EF \times ED / BW \times AT, \quad (1)$$

where CDI is the ingestion dose from drinking water (mg/kg/day) and C is the estimated concentration of pollutants in groundwater samples (mg/l). In this research, BW data were obtained from a article about the median body weights for different age groups of the Indonesian population [15]. The parameter values used to calculate CDI can be seen in Table 1.

Table 1. Calculation of parameters included in the CDI for health risk assessment

Parameter	Meaning	Value			Unit
		Adult	Children	Infant	
IR	Amount of drinking water	2,09[a]	1[b]	0.3[c]	L/day
EF	Exposure frequency	350[b]	350[b]	350[b]	days/year
ED	Exposure duration	30[b]	12[d]	1[d]	Year
BW	Body weight		According to age[e]		Kg
AT	Exposure time (lifetime)	23,360[d]	4,380[d]	365[d]	Days

[a] These data come from questionnaires.
[b] These data come from [14].
[c] These data come from [16].
[d] These data come from [17].
[e] These data come from [15].

IOP Conf. Series: Earth and Environmental Science **633** (2021) 012015 doi:10.1088/1755-1315/633/1/012015

The ingested amount of risk agent, which is calculated using Eq. 1, can be used to estimate the ingested HQ (hazard quotient). The HQ describes the risk of reporting oral non-carcinogens due to the ingestion of water, which is expressed as follows:

$$HQ = CDI / RfD \quad (2)$$

If HQ < 1, there is no apparent non-carcinogenic risk or an acceptable level of risk. If HQ > 1, there is an unacceptable risk of adverse non-carcinogenic effects [14]. RfD refers to the reference dose for ingestion (mg/kg.day), and it can be obtained from the Integrated Risk Information System (IRIS) database [14]. The RfD values used in this study are presented in Table 2.

Table 2. Toxicity responses (dose response) to oral exposure

Parameter	RfD (mg/kg.day)
NO_2	0.1
NO_3	1.6
Cr	0.003
Zn	0.3
Fe	0.7
Mn	0.14
Hg	0.0001
Cd	0.0005

3. Result and discussion

3.1 General characteristics of respondents

The respondents consisted of children to adults of various ages, ranging from 0–90 years. Age affects the body's resistance to disease and exposure to poisons/pollutants. In this study, the average age of the respondents was 29. However, those at the minimum and maximum ages—0 years and 90 years, respectively—are vulnerable and at risk of exposure to poisons.

The respondents were divided into three age groups: adults, children, and infants. The largest group is the adult population aged 17–90 years, with a percentage of 68.51%. The proportion of respondents aged 3–16 years was 24.62%, while the number of infants aged 0–2 years was 6.87%. The

3.2 Quality of groundwater samples

In total, 24 groundwater samples from 24 wells of Sumur Batu village residents were checked in situ to determine their pH and TDS values. Then, nitrite (NO_2), nitrate (NO_3), and heavy metal measurement was performed in the laboratory using a spectrophotometer, ICP-MS, and AAS technology. The test results for groundwater sample parameters in this study are presented in Table 3.

Table 3. Water quality test results

Sample	pH	TDS (ppm)	NO_2-N (ppm)	NO_3-N (ppm)	Cr (ppm)	Zn (ppm)	Fe (ppm)	Mn (ppm)	Hg (ppm)	Cd (ppm)
PP No. 82/2001	6–9	1000	0.06	10	0.05	0.05	0.3	0.1	0.001	0.01
1	4.8	143	0.61	2.94	<0.01	<0.01	<0.02	<0.01	<0.0006	<0.001
2	5.7	152	0.00	2.94	<0.01	<0.01	<0.02	<0.01	<0.0006	<0.001
3	4.7	104	0.61	2.53	<0.01	<0.01	<0.02	<0.01	<0.0006	<0.001
4	5.7	165	0.91	0.86	<0.01	<0.01	<0.02	<0.01	<0.0006	<0.001
5	5.5	78	0.30	1.63	<0.01	<0.01	<0.02	<0.01	<0.0006	<0.001
6	5.1	144	1.22	1.85	<0.01	<0.01	<0.02	<0.01	<0.0006	<0.001
7	5.8	90	0.30	1.26	<0.01	<0.01	<0.02	<0.01	<0.0006	<0.001
8	5.7	146	0.61	1.90	<0.01	<0.01	<0.02	<0.01	<0.0006	<0.001
9	7.1	151	0.30	0.27	<0.01	<0.01	<0.02	<0.01	<0.0006	<0.001

IOP Conf. Series: Earth and Environmental Science **633** (2021) 012015 doi:10.1088/1755-1315/633/1/012015

10	6.2	149	2.13	2.78	<0.01	<0.01	<0.02	<0.01	<0.0006	<0.001
11	5.6	142	0.61	2.44	<0.01	<0.01	<0.02	<0.01	<0.0006	<0.001
12	5.1	123	0.30	3.21	<0.01	<0.01	<0.02	<0.01	<0.0006	<0.001
13	5.9	193	0.91	1.74	<0.01	<0.01	<0.02	<0.01	<0.0006	<0.001
14	6.0	117	0.61	0.63	<0.01	<0.01	<0.02	<0.01	<0.0006	<0.001
15	6.1	413	0.61	2.48	<0.01	<0.01	<0.02	<0.01	<0.0006	<0.001
16	5.3	138	0.91	2.60	<0.01	<0.01	<0.02	<0.01	<0.0006	<0.001
17	5.0	97	0.61	1.47	<0.01	<0.01	<0.02	<0.01	<0.0006	<0.001
18	4.4	83	0.61	1.63	<0.01	<0.01	<0.02	<0.01	<0.0006	<0.001
19	5.6	288	0.91	6.01	<0.01	<0.01	<0.02	<0.01	<0.0006	<0.001
20	5.2	160	0.91	4.27	<0.01	<0.01	<0.02	<0.01	<0.0006	<0.001
21	5.4	83	0.30	2.57	<0.01	<0.01	<0.02	<0.01	<0.0006	<0.001
22	5.8	112	0.30	1.49	<0.01	<0.01	<0.02	<0.01	<0.0006	<0.001
23	5.7	144	0.30	2.33	<0.01	<0.01	<0.02	<0.01	<0.0006	<0.001
24	6.3	141	0.30	1.15	<0.01	<0.01	<0.02	<0.01	<0.0006	<0.001
Mean	5.6	148.2	0.6	2.2						
Std Dev	0.6	71.2	0.4	1.2						
Limit of Detection					**<0.01**	**<0.01**	**<0.02**	**<0.01**	**<0.0006**	**<0.001**

Based on Table 3, the 24 samples have varying pH values but tend to be acidic. Based on the Indonesian Government Regulation No. 82, enacted in 2001, only 4 of the 24 groundwater samples met the pH requirements for raw drinking water. Based on data from the Bekasi City government, the land around the Bantar Gebang village is alluvial. Alluvial soils have pH conditions that tend to be acidic, which can affect the pH of groundwater. Water with a low pH (below 6) can cause corrosion of metals (water pipes), dissolving the elements of Pb, Cu, and even Cd. Water acidity levels can trigger health problems, such as flu, allergies, and skin disorders, and if corrosion occurs, they can cause poisoning [18]. However, the TDS value of all water samples is considered to be safe according to Indonesian Government Regulation No. 82.

Of the 24 well water samples, only 1 has an NO_2-N content that meets the quality standard for drinking water. For NO_3-N content, all samples meet the quality standard. On average, the NO_2-N content is lower than the NO_3-N content. This shows that the process of nitrifying microorganisms in the soil is going well. NO_2 is usually present in groundwater at very low concentrations because it is very easily converted into NO_3 [19]. But even in low concentrations, the presence of nitrates in a water environment must be monitored because of their high toxicity. The Cr, Zn, Fe, Mn, Hg, and Cd contents in the 24 groundwater samples were very low and below the detection limit. This might be due to the groundwater flow from the landfill area, which does not lead to the Sumur Batu sub-district. According to Syamsu Rosid et al. (2011), who used the self-potential method to conduct their research, the subsurface groundwater flows in the Bantar Gebang landfill area are oriented from south to north [12]. However, in this study, the location of the study was to the east of the landfill area.

3.3 Health risk assessment

Laboratory test results for 24 samples showed heavy metal values below the detection limit. Based on the US EPA, when pollutant values are below the detection limit, the health risks can be calculated using half of the detection limit. The detection limit is the lowest concentration (not 0) below the level that can be measured with acceptable precision. At the detection limit, analytes are proven to exist, but the reported concentrations are estimates [20]. In this study, the non-cancer risk of drinking water was calculated in three age groups for real time, real time + 5, real time + 10, real time + 15, and lifetime. The real time risk is the risk after a long period of exposure before this research. The real time risk + 5 refers to the risk 5 years after the study was conducted, and so on. Lifetime risk refers to the risk if the resident continuously consumes polluted groundwater.

Table 4. Risk analysis of groundwater contaminated with heavy metals as a source of drinking water

Parameter		Percentage of respondents with HQ > 1 (%)					
		Real time	Real time + 5	Real time + 10	Real time + 15	Lifetime	Risk
Infant (n = 41)	NO_2	19.5	0	0	0	19.5	**At risk**
	NO_3	0	0	0	0	0	No risk
	Cr	0	0	0	0	0	No risk
	Zn	0	0	0	0	0	No risk
	Fe	0	0	0	0	0	No risk
	Mn	0	0	0	0	0	No risk
	Hg	0	0	0	0	0	No risk
	Cd	0	0	0	0	0	No risk
Child (n = 147)	NO_2	33.3	0	0	0	33.3	**At risk**
	NO_3	0	0	0	0	0	No risk
	Cr	0	0	0	0	0	No risk
	Zn	0	0	0	0	0	No risk
	Fe	0	0	0	0	0	No risk
	Mn	0	0	0	0	0	No risk
	Hg	0	0	0	0	0	No risk
	Cd	0	0	0	0	0	No risk
Adult (n = 409)	NO_2	0.2	0.2	1.2	1.7	0	**At risk**
	NO_3	0	0	0	0	0	No risk
	Cr	0	0	0	0	0	No risk
	Zn	0	0	0	0	0	No risk
	Fe	0	0	0	0	0	No risk
	Mn	0	0	0	0	0	No risk
	Hg	0	0	0	0	0	No risk
	Cd	0	0	0	0	0	No risk

Based on the estimated health risks calculated in Table 4, exposure to NO_3 and Cr, Zn, Fe, Mn, Hg, and Cd in groundwater samples incur no or limited risk in all age groups until 15 years after this study. However, there are risks associated with NO_2 in all age groups. As many as 19.5% of the infant age group are at risk of being affected by NO_2 through real time and lifetime exposure. In the child age group, as many as 33.3% of respondents are at risk of real time and lifetime NO_2 exposure. In the adult age group, there is a risk of NO_2 exposure at the real time, real time + 5, real time + 10, and real time + 15 time points. As many as 0.2% of the adult age group have a non-cancer health risk associated with NO_2 due to oral exposure in real time and the next 5 years; 1.2% has a risk of NO_2 exposure in the next 10 years; and 1.7% are at risk of NO_2 exposure in the next 15 years.

High nitrate exposure incurs a great risk, especially in infants and pregnant women [13]. In the bloodstream, nitrites oxidize iron in hemoglobin molecules and produce methemoglobin. This condition is called methemoglobinemia. Iron that has been oxidized can no longer bind oxygen, and the ability of blood cells to carry oxygen to the body is reduced, inhibiting the growth of organs and even leads to death. Also, years of low nitrate exposure can affect adult health by causing gastric cancer, respiratory disorders, headaches and fatigue, thyroid gland hypertrophy, and multiple sclerosis (an autoimmune disorder that attacks the central nervous system) [21].

Because Sumur Batu is directly bordered by the landfill area and the community is highly dependent on groundwater, the quality of groundwater in this area needs to be monitored. Especially among vulnerable infants, the consumption of groundwater for drinking needs to be limited. Breastfeeding until the age of two years is highly recommended, and the need for water for formula milk or complementary foods can be met using safer water like bottled water.

4. Conclusion

The contents of Cr, Zn, Fe, Mn, Hg, and Cd are below the limit of detection in all groundwater samples. This may be due to the direction of groundwater flow from the landfill, which does not lead to Sumur Batu. The results of a health risk analysis of NO_3 and heavy metals in groundwater in the Sumur Batu sub-district indicate that using this groundwater as a source of drinking water does not cause significant health effects in all age groups until 15 years later. However, there are risks associated with NO_2 in all age groups. Because the region is directly bordered by the landfill area and is highly dependent on groundwater, the quality of groundwater in this area needs to be monitored. Especially among vulnerable infants, the intake of groundwater needs to be limited.

5. References

[1] L. Xing, H. Guo and Y. Zhan, "Groundwater hydrochemical characteristics and processes along flowpaths in the North China," *J. of Asian Earth Sciences,* Vols. 70-71, pp. 250-264, 2013.

[2] G. Zhang, Y. Jiao and D. Lee, "A lab-scale anoxic/axic-bioelectrochemical reactor for leachate treatments," *Bioresour Technol,* vol. 186, pp. 97-105, 2015.

[3] K. Ravindra and S. Mor, "Distribution and health risk assessment of arsenic and selected heavy metals in Groundwater of Candigarh," *Environmental Pollution,* vol. 250, pp. 820-830, 2019.

[4] Bappeda Kota Bekasi, " Rencana pembangunan jangka menengah daerah Kota Bekasi 2018 – 2023," Bekasi, 2018.

[5] DLH Jakarta, "Portal Resmi UP Sampah Terpadu Dinas Lingkungan Hidup Prov. DKI Jakarta," 2019. [Online]. Available: https://upst.dlh.jakarta.go.id/tpst/index. [Accessed 20 April 2020].

[6] Liputan6, "Program Khusus : Mari Tinggalkan Pengolahan Sampah Primitif," 13 February 2004. [Online]. Available: https://www.liputan6.com/news/read/72093/mari-tinggalkan-pengolahan-sampah-primitif. [Accessed 23 12 2019].

[7] Tempo.co, "Metro : Kesia-siaan di TPST Bojong," 23 November 2004. [Online]. Available: https://metro.tempo.co/read/51447/kesia-siaan-di-tpst-bojong. [Accessed 23 12 2019].

[8] Pemerintah Kota Bekasi, "Bekasi Kota," 2019. [Online]. Available: https://www.bekasikota.go.id/pages/kondisi-geografis-wilayah-kota-bekasi. [Accessed 2019].

[9] T. Hutabarat and E. R. Pujiindiyati, "Analisis Polutan Logam Berat dalam Air pada Tempat Pembuangan Akhir Bantar Gebang, Bekasi dan Sekitarnya dengan Metode Analisis Aktivasi Neutron Instrumental," *ISSN,* pp. 2085-2797, 2010.

[10] B. Carolina, "Penentuan Status Mutu Air dengan Sistem STORET di Kecamatan Bantar Gebang," *Jurnal Geologi Indonesia,* vol. 2, no. 2, pp. 113-118, 2007.

[11] Indrayani E, "Analisis Kandungan Nitrogen, Fosfor dan Karbon Organik di Danau Sentani Papua," *Jurnal Manusia dan Lingkungan 22(2),* pp. 217-225, 2015.

[12] S. Rosid, R. N. Koesnodo and P. Nuridianto, "Estimasi aliran air lindi TPA Bantar Gebang Bekasi menggunakan metode sp," *Jurnal Fisika,* vol. 1, November 2011.

[13] M. Hunter H. Comly, "Landmark article 8 September 1945: Cyanosis in Infants Caused by Nitrates in Well Water," *JAMA,* vol. 257(20), pp. 2788-2792, 1987.

[14] USEPA, "Risk assessment guidance for superfund: process for superfund: process for conducting probabilistic risk assessment (volume III-part a, 540-R-502-002)," 2001. [Online]. [Accessed 12 2019].

[15] S. Muljati, A. Triwinarto, N. Utami and Hermina, "description of median number of weight and height classified by age group on healthy indonesian citizens based on riskesdas 2013 result," *Penelitian Gizi dan Makanan,* vol. 39, no. 2, pp. 137-144, 2016.

[16] C. P. Ahada and S. Suthar, "Groundwater nitrate contamination and associated human health risk assessment in southern districts of Punjab, India," *Environmental Science and Pollution*

Research, vol. 25, pp. 25336-25347, 2018.

[17] N. Adimalla and H. Qian, "Groundwater quality evaluation using water quality index (WQI) for drinking purposes and human health risk assessment in an agricultular region of Nanganur, south India," *Ecotoxicological and Environmental Safety,* vol. 176, pp. 153-161, 2019.

[18] A. Herlambang, "Pencemaran Air dan Penanggulangannya," *J. Air Indonesia,* vol.2, no. 1, 2006.

[19] E. Elisante and A. Muzuka, "Assessment of Sources and Transformation of Nitrate in Groundwater on the Slopes of Mount Meru, Tanzania," *Environ. Earth Sci.,* vol. 75(3), pp. 1-15, 2016.

[20] US Environmental Protection Agency, "Chemical Concentration Data Near The Detection Limit," in *Hazardous Waste Management*, Philadelphia, 1991.

[21] P. Gatsetya and M. Agrirova, "High nitrate levels in drinking water may be a risk factor for thyroid disfunction in children and pregnant women living in rural Bulgarian areas," *Int J. Hyg Environ Health,* vol. 211, pp. 555-559, 2008.

Acknowledgments

This research was funded by the Water for Women Fund, given by the Australian Government Department of Foreign Affairs and Trade (no. WRA1: 1004). It was also funded by PTUPT 2 (ADD-NKB-2943/UN2.RST/HKP.05.00/2020), given by the Ministry of Research, Technology, and Higher Education of the Republic of Indonesia. Thank you to Juliet Willetts, Tim Foster, and Franzisca Genter from the University of Technology Sydney, and also Angela Harris from North Carolina State University, for their willingness to share knowledge and extraordinary collaboration.

IOP Conf. Series: Earth and Environmental Science **633** (2021) 012016 doi:10.1088/1755-1315/633/1/012016

Effect of boiling and water storage practices on *E. coli* contamination of drinking water in the city of Bekasi (case study: Jatiluhur, Sumur Batu, and Jatirangga Villages)

I Imtiyaz[1], G L Putri[2], D M Hartono[1], F Zulkarnain[3], C R Priadi[1,2*]

1Faculty of Engineering, Civil and Environmental Engineering Department, Universitas Indonesia, Kampus UI Depok, 16424, Indonesia
2 Center for Sustainable Infrastructure Development, Universitas Indonesia, Kampus UI Depok, 16424, Indonesia
3 Faculty of Mathematics and Natural Science, Geography Department, Universitas Indonesia, Kampus UI Depok, 16424, Indonesia
E-mail: crpriadi@ui.ac.id

Abstract. In the city of Bekasi, a large proportion of the population uses groundwater and is at risk of *Escherichia coli* contamination. Water can remain contaminated by *E. coli* even after treatment, and it may be recontaminated through storage practices. This research aims to analyze the effect of boiling and the correlation of water storage practices to *E. coli* contamination in the drinking water of households in the villages of Jatiluhur, Sumur Batu, and Jatirangga. Among 54 randomly surveyed households, 98.1% boiled their water before drinking. The results show that 67% showed decreased *E. coli* after boiling. Boiled water with a low risk of *E. coli* made up 64.8% of the samples, a medium risk made up 25.9%, a high risk made up 7.4%, and a very high risk made up 1.9%. Observation of water storage practices showed that 51% of households store water in a jug after treatment, and 35.3% store it in a kettle or pot. Residents used a container equipped with a lid 94.1% of the time. The statistical analysis revealed no correlation between water storage practices and *E. coli* contamination in drinking water. Boiling water can be a fairly effective way to decrease *E. coli* contamination; however, other factors that can recontaminate water, such as storage hygiene and hand washing, require attention in future studies.

1. Introduction

Water is the most essential element for humans and other living things [1]. Humans need water, especially for drinking. Based on the Ministry of Health Regulation of the Republic of Indonesia Number 492 of 2010, drinking water is water that may or may not undergo treatment that meets health requirements and can be drunk directly. Water for drinking must be free from contamination and safe for consumption. In accordance with Sustainable Development Goals (SDG's) 6.1 target, water consumed by the community should be easy to access and free of contaminants. One indicator of water contamination is the presence of *Escherichia coli*. Ministry of Health Regulation 492 of 2010 states that the microbial quality standard for *E. coli* in drinking water should be 0 in 100 ml samples [2].

E. coli are commensal bacteria, intestinal pathogens, and extraintestinal pathogens that can cause urinary tract infections, meningitis, and septicemia. Pathogenic *E. coli* can cause diarrhea in humans [3]. In Indonesia, the level of diarrhea in infants reached 12.3% in 2018 [4]. In fact, during the same year, this disease caused 4.76% of deaths in children in Indonesia [5]. Diarrhea can also cause stunting or chronic malnutrition problems due to lack of nutrition over a long period, resulting in growth disorders, or stunting, in children. In 2019, the percentage of stunting in children under five years old in Indonesia was 27.67% [6].

Diarrhea due to *E. Coli* contamination is associated with both sanitation conditions and inadequate access to sources of clean drinking water, which becomes a pathway for organisms to enter the human body [7]. Water treatment and hygienic storage can decrease the concentration of *E. coli.*

Individual households can treat water to ensure its safety through a variety of methods, such as boiling, filtration, chemical treatment, and solar disinfection [8]. Boiling is perhaps the oldest way to disinfect water and is widely used at the household level. If it is practiced properly, boiling can be one of the most effective ways to kill or deactivate the pathogens that can be transmitted through water, including bacterial spores and protozoa [9]. In Indonesia, boiling water is an easy, practical, and inexpensive way to improve water quality [10]. Boiling drinking water has become a culture for some people in Indonesia. Studies conducted at the Cikapundung riverbank area in Bandung showed that 62% of the users of protected wells treated their drinking water by boiling, and 67% of the users of unprotected wells boiled their water before drinking [11]. Boiling water can reduce the average concentration of *E. coli* by 83%; however, *E. coli* was still detected in 70.4% of the water of those who reported that they boiled their water [12]. This shows the that water is not completely free from *E. coli* contamination even after treatment.

Another factor that can affect the presence of *E. coli* after boiling is the use of water storage containers. The type of container used and the availability of a lid are the aspects that should be taken into consideration [13]. When water is stored in gallons, bottles, or thermoses, it has a 42% chance of contamination, and when it is stored in dispensers, it has a 37% chance of contamination with *E. coli* [12]. In addition to these aspects, it is important to keep the storage container clean by washing it at least once a week or more often if possible [14].

Bekasi is the most populated city in the province of West Java; it has a population of 2,873,484 people. The population growth rate reached 2.64% from 2013 to 2017. The city is also densely populated, with an average density of 13,651.4 people/km^2 [15]. Unfortunately, the high population does not have corresponding adequate water services. In Bekasi, the Regional Drinking Water Company (PDAM) supplies piped water services, but it services only 26.8% of the population. Meanwhile, the rest of the population uses groundwater to fulfill their water needs; 2.1 million people construct boreholes or dig wells at their dwellings, in their yards, or in plots to access groundwater for their household [15]. Considering the large number of people using groundwater in Bekasi and the effect of *E. coli* contamination on drinking water, it is important to understand its extent. As mentioned above, *E. coli* is still detected after boiling, and usage of water containers can affect *E. coli* contamination. Therefore, this study will analyze the effects of boiling water and water storage practices for drinking water in Bekasi and, especially, in three urban villages: Jatiluhur, Sumur Batu, and Jatirangga.

2. Materials and methods

2.1. Study location

The study locations for this research are three urban villages in Bekasi: Jatiluhur, Sumur Batu, and Jatirangga. The study was conducted during the rainy season, from February to March 2020. These study locations were chosen for their high population density, high poverty rates, and lack of piped water connections or PDAM, which results in a high dependence on groundwater use.

Out of the 98 households sampled in this study, only 54 used groundwater as their source of drinking water (instead of bottled water, gallon water, or refill water). Since this research is focused on boiling water for drinking, and people do not boil their bottled water, gallon water, or refill water, only data from the 54 households were analyzed. The study locations are shown in Figure 1 below.

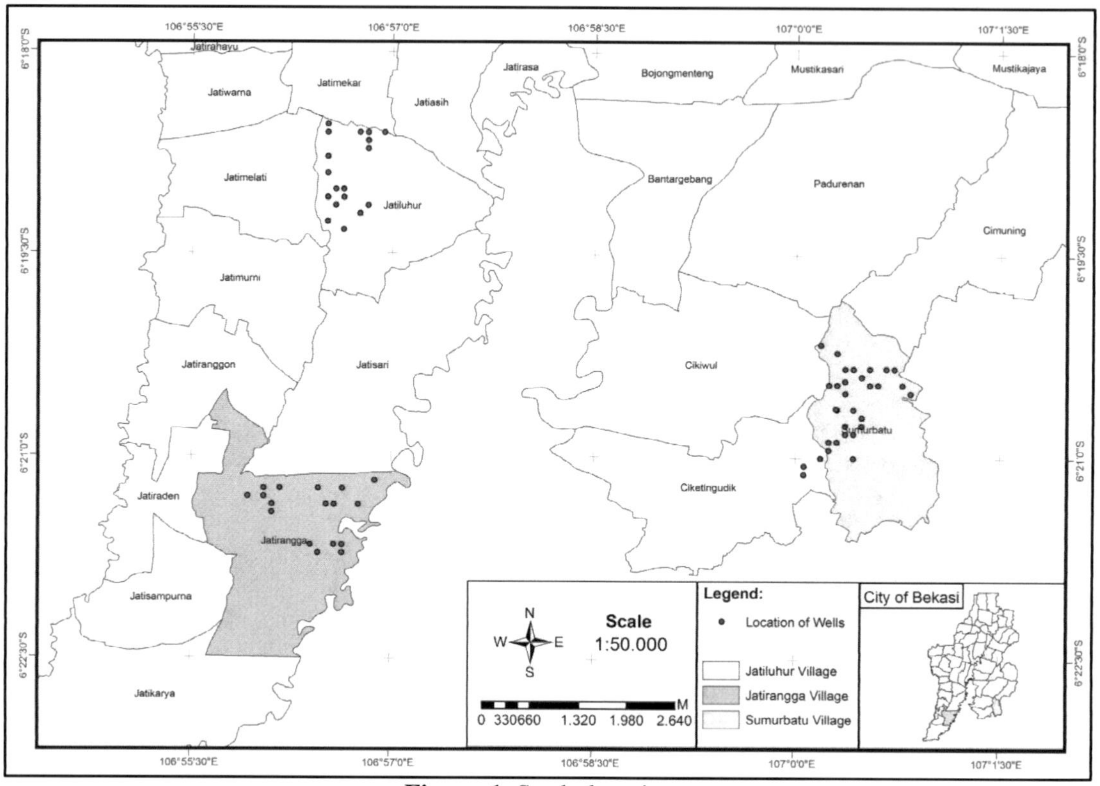

Figure 1. Study location map

2.2. Data collection

Data collection was carried out using both a questionnaire that included questions regarding how the households treat groundwater before drinking it and observations of their storage containers. The Survey Solution application was used to facilitate data collection for the interviewer. Coordinate data from the location of the research well with GPS was also collected to be mapped using the ArcGIS application. In addition to collecting data using questionnaires, water sampling was done in the groundwater source (boreholes and/or dug wells) and at the point of use (POU) for drinking water. About 100 ml of each water sample was collected in a sterile Whirl-Pak and stored in a cooler bag with a temperature of ≤4°C. The samples were taken to the laboratory and were measured for *E. coli* concentration per 100 mL using IDEXX-Colilert 18. This method uses a Quanti-tray with 49 large columns and 48 small columns to see the color changes that occur after incubation with reagents [16],[10]. The maximum storage time before testing was six hours. The *E. coli* concentration was calculated using the most probable number (MPN). To control the quality of the water samples and to ensure that they are safe from contamination, field blanks, laboratory blanks, and duplicates were taken once every day.

3. Results and discussion

3.1. Effectiveness of boiling water on E. coli reduction

Based on the results, households in the study location treat their groundwater by boiling it and adding chlorine to it. Data shows that 98.1% of the total samples (53 households) treated their water by boiling, and the remaining 1.89% (1 household) treated their water by adding chlorine. The majority of the population treated their water by boiling it prior to drinking.

This study also found that boiling water did not remove *E. coli* completely. Comparing the concentration of *E. coli* in the groundwater source and the concentration of *E. coli* in the drinking water after boiling revealed that *E. coli* was only removed from 67% of the samples. These results prove that boiling is a fairly effective way to reduce the concentration of *E. coli*. While boiling decreases *E. coli* levels, it can still be detected in drinking water [12]. However, this study also found that the concentration of *E. coli* did not change in 19% of the samples and that it increased in 14% of the samples, as shown in Figure 2 below. It is possible that this is due to inadequate boiling procedures, improper water handling after boiling (leading to recontamination), or false reports of boiling [17]. Moreover, when water begins to cool after boiling, it is vulnerable to recontamination from hands and utensils, because it contains no residual disinfectant and is often stored in open vessels [9].

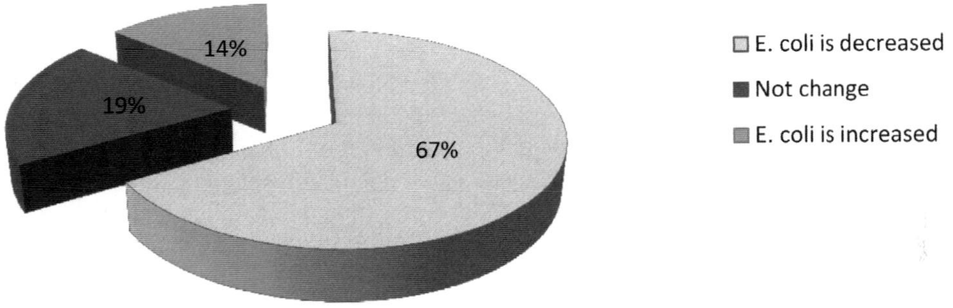

Figure 2. Effectiveness of decreasing *E. coli* by boiling

3.2. Risk assessment of E. coli after boiling

Although boiling water can be a fairly effective way to decrease the contamination of *E. coli*, *E. coli* remains after treatment. A risk assessment of *E. coli* is needed to determine whether water is safe for consumption. Assessing the risk of drinking water supplies can increase trust in drinking water safety [18]. Based on WHO's 2011 report, "Guidelines for Drinking-Water Quality", the risk level of *E. coli* can be divided into four categories: low, medium, high, and very high risk, which are defined as *E. coli* levels of <1, 1–10, 11–100, and >100 MPN/100ml, respectively [19]. The results showed that 64.8% of boiled water had low risk, 25.9% had medium risk, 7.4% had high risk, and 1.9% had very high risk *E Coli*. contamination (Figure 3). This result indicates that drinking water has the potential to cause waterborne disease. A prior study in Aceh province, Indonesia showed that boiling water does not guarantee that it will be free from *E. coli* contamination [17]. Also, a study in Tanzania found that households that boiled drinking water had higher percentages of diarrhea (16%) than control households (9%) [7]. Therefore, it is important to discover the source of recontamination with *E. coli* after boiling in order to address the health issue caused by *E. coli* contamination. As mentioned before, water storage practices may contribute to increased *E. coli*. Observation of water storage practices by the households in the study locations is explained in the next part.

IOP Conf. Series: Earth and Environmental Science **633** (2021) 012016 doi:10.1088/1755-1315/633/1/012016

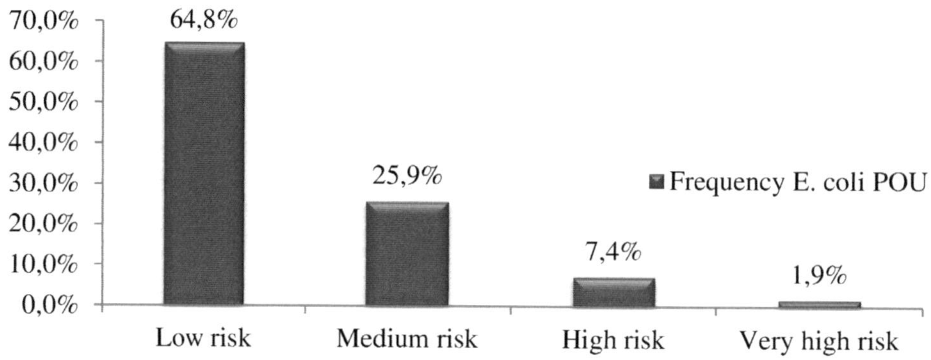

Figure 3. Groundwater treatment

3.3. Correlation of water storage practices with E. coli contamination
Data for water storage practices was processed for 51 out of the 54 samples and remained unknown for the other 3 samples. Based on the data, 51% of households (n=51) store boiled water in a jug, 35.3% use a kettle/teapot, and the rest use a pot, gallon, bottle, or bucket (Table 1). The households that use a cover on their water container made up 94.1% of the sample (Table 1). Covering the water container is important to protect the water from contamination. If contamination occurs, organic compounds that remained after boiling can potentially support bacterial growth [8]. Based on previous studies, about 44% of boiled water can be contaminated with *E. coli* because the container is open and easily touched by hands [10].
In addition, other factors that influence the risk of contamination are higher temperatures, length of storage time, and the presence of air or dust particulates [13]. The data shows that 39.2% of the samples kept their water for one day, and 41.2% kept their water after boiled one day ago (Table 1). People tend to boil only the quantity of water they need for a short period, so that the water does not need to be stored for a long time after boiling.

Table 1. Water storage containers (n=51)

		n	%
Type of Container	Jug	26	51.0%
	Kettle/Teapot	18	35.3%
	Bottle	1	2.0%
	Bucket	1	2.0%
	Gallon	2	3.9%
	Pot	3	5.9%
Container Cover	Yes	48	94.1%
	No	3	5.9%
Storage Time in Container (days)	0	16	31.4%
	1	20	39.2%
	2	9	17.6%
	3	5	9.8%
	7	1	2.0%
Days After Water Undergo Boiling (days)	0	17	33.3%
	1	21	41.2%
	2	9	17.6%
	3	3	5.9%
	7	1	2.0%

IOP Conf. Series: Earth and Environmental Science **633** (2021) 012016 doi:10.1088/1755-1315/633/1/012016

The correlation of *E. coli* contamination at POUs with storage containers was analyzed using the SPSS application and Spearman's correlation. Spearman's test is used to find the compatibility between the dependent variable and the independent variable using the ordinal data scale, which is changed to rank hypothesis testing using two tails to test whether the sample is larger or smaller than the range of values on both sides of the specified data range. The concentration of *E. coli* is divided into three categories: ≤1 MPN/100 ml, ≤10 MPN/100 ml, and ≤100 MPN/100 ml. The results showing correlation between the water quality at POUs and storage practices are shown in Table 2.

Table 2. Correlation of *E. coli* at point of use with storage practices (n=51)

	≤1 MPN/100 ml		**≤100 MPN/100 ml**		**≤100MPN/100 ml**	
	Correlation Coefficient	***Sig. (two-tailed)***	***Correlation Coefficient***	***Sig. (two-tailed)***	***Correlation Coefficient***	***Sig. (two-tailed)***
Type of Water Storage Containers	-0.034	0.814	0.101	0.480	0.090	0.530
Container Cover	-0.146	0.306	-0.082	0.565	-0.035	0.805
Storage Time in the Container (days)	0.121	0.399	0.141	0.322	-0.005	0.972
Days After Water Undergo Boiling (days)	0.101	0.483	0.223	0.115	-0.020	0.887

Based on the results of Spearman's test, none of the variables are significant (>0.05). This indicates that there is no correlation between storage containers, container covers, storage time, or length of time after processing with *E. coli* contamination at POUs water. Furthermore, the output results above the correlation coefficient values are below number 0.25 entirely, revealed that the coefficient values indicate a very weak correlation between the independent and dependent variables [20]. The results confirm previous studies, which found no relationship between improved storage containers, storage time with *E. coli* contamination [7],[17]. That might be because the parameters of container hygiene and hand washing were not observed and included. Container hygiene is important and can be maintained by cleaning at least once a week or more often if possible [14]. Pouring water is also preferable to scooping it from a glass container because the water can be easily contaminated when the user's hands are dirty [14]. Based on previous studies, if water has been contaminated from water sources and improper boiling practices was done, then the type of water storage container will have no effect on protection from microbial contamination [17].

4. Conclusions

Boiling water is a common practice in almost all households at the study site. As many as 98.1% of households treat their groundwater by boiling. After boiling, *E. coli* was decreased in 67% of the samples, no change was detected in 19% of the samples, and *E. coli* was increased in 14% of the samples. A total of 64.8% of boiled water showed a low risk of *E. coli*, 25.9% showed a medium risk of *E. coli*, 7.4% showed a high risk of *E. coli*, and 1.9% showed a very high risk of *E. coli*. Most of the households (51%) store their boiled water in a jug, followed by 35.3%, who store it in a kettle or teapot. In addition, 94.1% of households used water containers with a lid or cover, and most of them stored the boiled water for only one day before consuming it. The statistical analysis showed no correlation between the water storage practices (type of water container, availability of container cover, storage time in the container, days since boiling) and *E. coli* contamination at the POU for drinking. Other parameters, such as container hygiene and hand washing, are worth considering for future research.

5. References

[1] Sutopo M F Sanim B Syaukat Y and Mawardi M I 2011 Environmental Engineering Journal vol 12 *Analysis of Willingness to Pay for Environmental Services in Integrated Water Resources Management in Indonesia (Case Study of Cisadane Hulu Watershed)* (Jakarta: BPPT) p 17–23

[2] Ministry of Health 2010 *Regulation of the Minister of Health of the Republic of Indonesia Number 492/MENKES/PER/IV/2010 Regarding Drinking Water Quality Requirements* (Indonesia: Ministry of Health)

[3] Halim F Warouw S M Rampengan N H and Salendu P 2017 Sari Pediatri vol 19(2) *Correlation between the Colonies Count of Escherichia Coli with The Grade of Dehydration in Acute Diarrhe* (Manado: Sari Pediatri) pp 81–85

[4] Ministry of Health 2018 *Main Riskesdas Results* (Jakarta: Ministry of Health)

[5] Budijanto D 2018 *Indonesian Healthy Profile* (Jakarta: Ministry of Health)

[6] Izwardy D 2020 *Study of Integrated Susenas Toddler Nutrition Status 2019* (Jakarta: National Institute of Health Research and Development)

[7] Mattioli M C Boehm A B Davis J Harris A R Mrisho M and Pickering A J 2014 PLoS One vol 9(1) *Enteric Pathogens in Stored Drinking Water and on Caregiver's Hands in Tanzanian Households with and without Reported Cases of Child Diarrhea* (San Francisco: Public Library of Science) pp 1–33

[8] Luby S P Syed A H Atiullah N Faizan M K and Fisher-Hoch S 2000 Int. J. Inf. Diseases vol 4(1) *Limited Effectiveness of Home Drinking Water Purification Efforts in Karachi, Pakistan* (Pakistan: The Aga Khan University)

[9] Clasen T McLaughlin C Nayaar N Boisson S Gupta R Desai D and Shah N 2008 Am. J. Trop. Med. Hyg. vol 79(3) *Microbiological Effectiveness and Cost of Disinfecting Water by Boiling in Semi-urban India* (London: The American Society of Tropical Medicine and Hygiene) pp 407–413

[10] Sodha S V Menon M Trivedi K Ati A Figueroa M E Ainslie R Wannemuehler K and Quick R 2011 *Microbiologic Effectiveness of Boiling and Safe Water Storage in South Sulawesi, Indonesia* (City: IWA Publishing) p 577–85

[11] Raksanagara A S Santanu A M Sari Y I S Sunjaya D K Arya I F D Agustian D 2017 MKB vol 49(2) *Integrated Behavior Model: Factors Influencing Clean Water Use among Urban Slum Dwellers* (Bandung: Universitas Padjadjaran) p 122–31

[12] Cronin A A Odagiri M Arsyad B Nuryetty M T Amannullah G Santoso H Darundiyah K Nasution N 'A 2017 Int. J. Hyg. Environ. Health *Piloting Water Quality Testing Coupled With A National Socioeconomic Survey in Yogyakarta Province, Indonesia, Towards Tracking of Sustainable Development Goal 6* (Elsevier) p 11

[13] Sobsey M D 2002 *Managing Water in the Home: Accelerated Health Gains from Improved Water Supply* (Geneva: World Health Organization Press)

[14] Howard G 2002 *Water Quality Surveillance: A Practical Guide* (Leicestershire: WEDC, Loughborough University)

[15] Development Planning Agency at Sub-National Level Bekasi City 2018 *Bekasi City Regional Medium-Term Development Plan 2018–2023* (Bekasi: Bekasi City Government)

[16] Taonameso S Mudau L S Traoré A N and Potgieter N 2019 Water Supply Jurnal vol 19(1) *Borehole water: a potential health risk to rural communities in South Africa* (South Africa: Journal of Water Supply) pp 128–136

[17] Gupta S K Suantio A Gray A Widyastuti E Jain N and Rolos R 2007 Am. J. Trop. Med. Hyg. vol 76(6) *Factors Associated With E.Coli Contamination of Household Drinking Water Among*

IOP Conf. Series: Earth and Environmental Science **633** (2021) 012016 doi:10.1088/1755-1315/633/1/012016

Tsunami and Earthquake Survivors, Indonesia vol 76 no 6 (Atlanta: The American Society of Tropical Medicine and Hygiene) pp 1158–1162

[18] WHO 2006 *Guidelines for Drinking-Water Quality* (first addendum to the third edition) (Geneva: World Health Organization Press)

[19] WHO 2011 *Guidelines for Drinking-Water Quality* (fourth edition) (Geneva: World Health Organization Press)

[20] Machali I 2015 *Statistics Are Easy: Using SPSS as a Statistics Aid* (Yogyakarta: Lembaga Ladang Kata)

Acknowledgments

This research was funded by the Water for Women Fund, given by the Australian Government Department of Foreign Affairs and Trade (number WRA1: 1004). It was also funded by PTUPT (number: ADD-NKB-2942/UN2.RST/HKP.05.00/2020) given by the Ministry of Research, Technology, and Higher Education of the Republic of Indonesia. Thank you to Juliet Willetts, Tim Foster, Franziska Genter from the University of Technology Sydney and Angela Harris from North Carolina State University for their willingness to share knowledge and extraordinary collaboration.

IOP Conf. Series: Earth and Environmental Science **633** (2021) 012017 doi:10.1088/1755-1315/633/1/012017

Risk analysis of radioecological effect of e-waste lead (Pb) content on life forms

Emetere M.E.[1,2] , Aghogho E.[1] & Afolalu S.A. [3]

1Department of Physics, Covenant University Canaanland, P.M.B 1023, Ota, Nigeria.
2Department of Mechanical Engineering Science, University of Johannesburg, South Africa.
3Department of Mechanical Engineering, Covenant University Canaan land, Ota, Nigeria
E-mail: emetere@yahoo.com

Abstract: Electronic waste (e-waste) has been proven to pollute the soil, air, and groundwater significantly. Aside from the physical effect of coming in contact with pollution, the radioecological impact of the pollution may be dangerous either directly or indirectly to persons living at the source of pollution. In this study, the lead (Pb) pollution from e-waste was used to calculate several organisms' radioecological parameter. It is was observed that there was a moderate reduction in the number of plants with leaves (22%), a major reduction in cumulative stem growth (43% the control value), a severe increase in the number of chlorophyll mutations, severe changes in leaf colour (4.9-fold), and a major increase in mortality for all organism. It is recommended that landfill areas where e-wastes are dumped be excavated 9 m depth as a corrective measure to reduce the negative outcome on landfill workers.

1. Introduction

Due to the lack of finances available to individuals in developing countries, the stunted growth in technological advancement has been boosted the importation of used electronic equipment from developed countries. This continuous event has led to brokers and business start-ups springing up to channel used equipment from developed countries, leading to significant electronic waste (e-waste). Currently, electronic industries are known to be the swiftest growing industries. Electronic waste is continuously growing at a steady rate. Electronic waste can be classified as irreplaceable electronic equipment such as printers, cell phones, video gaming systems, television sets, faxes, and stereo.
Electronic waste comprises of toxic substances as follows: barium, mercury, lead, and cadmium. These harmful substances require management, and they contain useful recoverable resources. Studies have shown that electronic waste's various dangers pose both a threat to surrounding and human health. Electronic waste is among the largest and fastest-growing waste stream in developing communities [1]. The heavy metals present in electronic devices and products include mercury, lead, gallium, selenium, arsenic, zinc, cobalt, tin, palladium, and aluminum [2]. Hino et al. [3] reported the significant number of e-wastes in the form of personal computer (PC), that is, 40,000 tons were disposed of in 1999; 50,000 tons of PCs were disposed of in 2000; 80,000 tons of PCs were disposed of in 2002. If we perform a rough extrapolation of the tons of e-waste from Japan since 1999, it should be around two million tons. Some are most time-shifted to developing countries whose populace patronizes it as the

only way they could own their PCs [4]. Lead (Pb) has been reported to be more abundant during e-waste processing. This observation is because lead used in the CRT manufacture in the form of lead oxide is used to shield x-rays generated within the operating CRT to move outside the compartment [5].

Smith et al. [6] reported that lead (Pb) by weight has 8% of the content of cathode ray tubes (CRTs) in computer monitors and TVs. Waldir et al. [7] reported that printed-wiring boards (PCB) consist of approximately 26% metal, made up mainly of copper, lead, aluminum, iron and tin, and other heavy metals as cadmium and nickel. Pan et al. [8] reported that the production of printed circuit boards in China is growing at 8.7% a year, making China the second-largest producer globally. Most electronic devices contain printed circuit boards (PCBs), which are composed of metals (such as copper, aluminum, tin, lead, and precious metals (gold, silver, and platinum), ceramics, and polymeric materials [7]. Lead contamination is prevalent in most municipal landfill leachate [8]. Pb is a developmental neurotoxicant, and blood Pb levels of ≥5 μg/dL in early childhood are recognized as having adverse effects (CDC, 2012). Cadmium (Cd) is also an element of concern in e-waste management. It has a biological half-life of ~26 years, and this metal accumulates primarily in the kidneys [9-10].

E-wastes are regarded as one of the influential causes of cancer in previous years. That is why most scientists are working hard to advance the replacement of the leading electronic component with biodegradable components that are more efficient. Already several countries have placed restrictions elements (also known as Restriction of Hazardous Substances, (RoHS)) that cannot be used by Electronics Company. Directive 2002/95/EC of the European Union restricts certain hazardous substances in electrical and electronic products - lead, mercury, cadmium, hexavalent chromium, and flame retardants [11]. In this research, the radioactive implication of lead contaminants was carried-out using ERICA 1.3 software.

2. Methodology

In this research, the ERICA 1.3 software was used for assessing the radiological risk of lead pollution to terrestrial, freshwater, and marine biota. The lead (Pb) concentrations were obtained from the literature [7, 12-13] to estimate risks in animals and plants. The dose rates used were 40μGyh-1 and 10μGyh-1 at different alpha and beta emissions to examine the risk analysis at a point source. The radioecology dataset was obtained using the high points in Figure 1 to understand radionuclides' transfer mechanisms.

The dose rates are calculated using the equivalent dose for radiation Λ (which can be α, β, γ, X, n, etc.) using

$$H_{\Omega} = \Sigma_{\Lambda} w_{\Lambda} D_{\Omega\Lambda}$$

where w_{Λ} is the radiation weighting factor, $D_{\Omega\Lambda}$ is the absorbed radiation, and it is measured in Gy w_{Λ} allows the introduction of Sv as the unit for H_{Ω}. The total equivalent dose for the whole organism is therefore calculated as :

$$H = \Sigma_{\Omega} w_{\Omega} H_{\Omega},$$

The internal dose rates from lead and its isotope are presented in Table 1. The activity concentration in the organism is presented in Table 2.

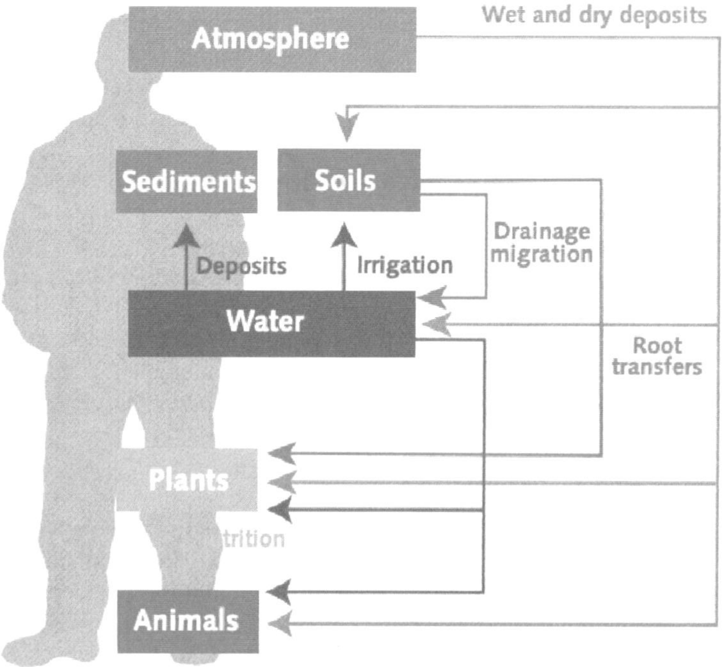

Figure 1. Modalities for calculating the radioecology (IPSN)

Table 1. Internal dose rate of lead and its isotope on plants and animals

Internal Dose Rate [µGy h-1] ----- Isotope	**Amphibian**	**Bird**	**Flying insects**	**Grasses & Herbs**	**Arthropod - detritivorous**	**Mammal - large**
Pb-210	0,011515392	0,0011067	0,000929628	0,0199716	0,0172482	0,022695
Po-210	0,2418	0,2418	0,2418	0,8742	0,8742	0,8742
Internal Dose Rate [µGy h-1] ----- Isotope	**Shrub**	**Mammal - small-burrowing**	**Mollusc - gastropod**	**Reptile**	**Tree**	**Lichen & Bryophytes**
Pb-210	0,0199716	0,0115056	0,00561	0,0217872	0,022695	0,000798883
Po-210	0,8742	0,465	0,465	0,07998	0,7998	0,183315946

Table 2. Activity concentration of lead and its isotope on the organism

Activity Concentration in the organism [Bq kg-1 f.w.] ----- Isotope	Amphibian	Bird	Flying insects	Grasses & Herbs	Arthropod - detritivorous	Mammal - large
Pb-210	47,04	4,34	4,34	89	89	89
Po-210	13	13	13	47	47	47
Activity Concentration in the organism [Bq kg-1 f.w.] ----- Isotope	**Shrub**	**Mammal - small-burrowing**	**Mollusc - gastropod**	**Reptile**	**Tree**	**Lichen & Bryophytes**
Pb-210	89	47	25	89	89	4,3
Po-210	47	25	25	4,3	43	10

3. Results and discussion

The activity concentration for lead and its isotope in soil and air is presented in Figure 2a as Pb (72%) and Po (28%). The total dose rate on each organism is presented in Figure 2b. Grasses and herbs, Arthropod, Mammal (human, etc.), and shrub had the highest total dose rates because they are directly exposed to both air and soil. Due to the height of an average tree, it receives lesser dose rates compared to shrubs.

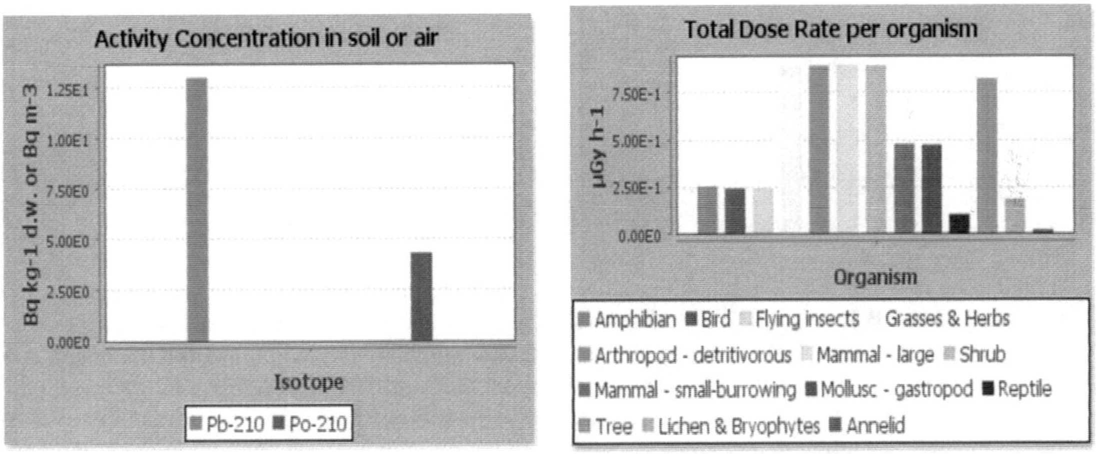

Figure 2. Radioecology effect (a) activity concentration in soil or air (b) total dose rate per organism

The dose rates used were 40µGyh-1 and 10µGyh-1 [14-16]. The dose rate screening value at 40µGyh-1 and 10µGyh-1 are presented in Figures 3a and 3b, respectively. At a lower dose rate (10µGyh-1), all organisms' screening values are uniform, as shown in Figure 3b. However, at a high dose rate (40µGyh-1), the screening value varies from one organism to another. Grasses and herbs, shrub, tree and lichen, and bryophytes had higher screening values indicating that plants have higher screening values than animals at high dose rates. The risk quotient for the categories of dose rates was presented in Figure 4.

IOP Conf. Series: Earth and Environmental Science **633** (2021) 012017 doi:10.1088/1755-1315/633/1/012017

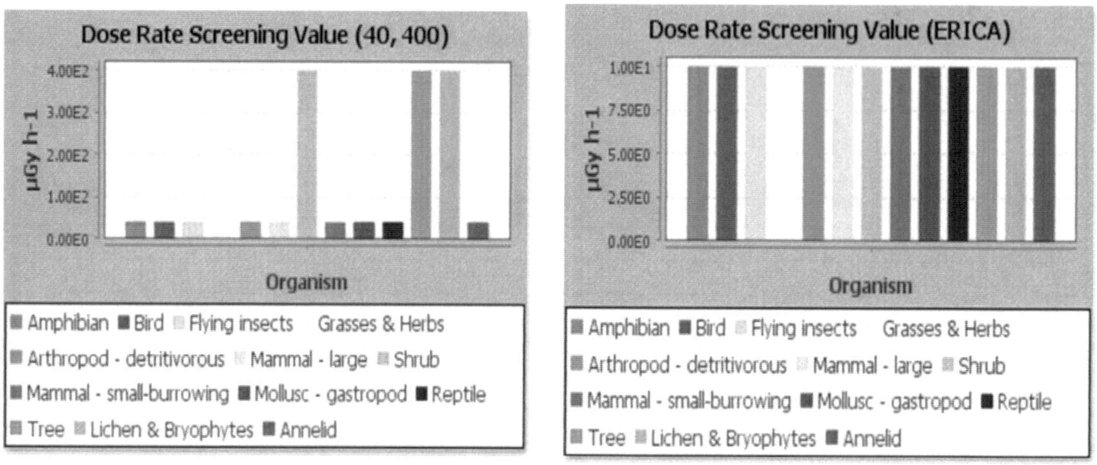

Figure 3. Dose rate screening value (a) dose rate = 40µGyh-1 (b) dose rate = 10µGyh-1

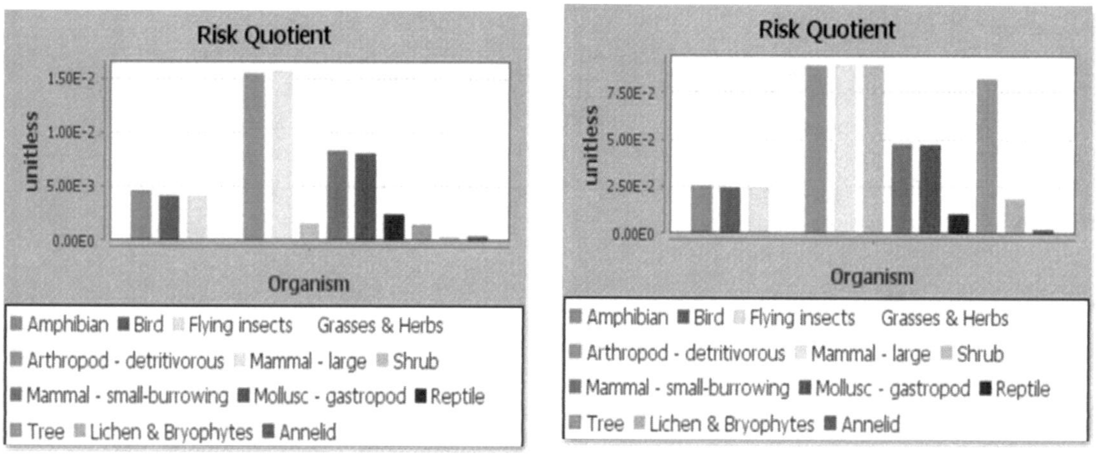

Figure 4. Risk quotient (a) dose rate = 40µGyh-1 (b) dose rate = 10µGyh-1

At a high dose rate (i.e., 40µGyh-1), the risk quotient can be used to classify the organism's reactions within the ecosystem. There are five types of risk quotient classification of Pb contamination. Class A, B, C, D, and E show the quotient risk in descending order. Class A comprises of the mammals (large) and arthropods; class B is the small burrowing mammals and mollusk (gastropod); class C is the amphibian, birds, and flying insects; class D is the grasses and herbs, shrubs, reptiles, and tries; class E is the lichen and bryophytes, and annelid. At low dose rate (i.e., 10µGyh-1), class A is grasses and herbs, arthropod, mammals (large), trees, and shrubs; class B is mammals and mollusk (gastropod); class C is amphibian, bird, and flying insect; class D is a reptile, and lichen and bryophytes; and class E is annelid. The risk quotient classification in low and high-risk shows that organisms' existence is threatened with realities such as the moderate reduction in the number of plants with leaves (22%), a significant reduction in cumulative stem growth (43% the control value), severe increase in the number of chlorophyll mutations, severe changes in leaf colour (4.9-fold), and a major increase in mortality for all organism.

4. Conclusion

Lead pollution of air, soil, and groundwater has been proven to go beyond mere physical contamination. The radioecology impact on organisms is enormous as it ultimately leads to death or eradicating salient organisms in the ecosystem. Since the volume of lead content in electronics is relatively high, that is, compared to other heavy metals. It is recommended that landfill areas where e-wastes are dumped be excavated 9 m depth. Slabs should be constructed in the excavated site before a new soil layer is refilled to mitigate organism extinction. The slab is to reduce the radioactive nature of heavy metals in e-waste.

5. References

[1] Pariatamby A; Victor D. (2013). Policy trends of e- waste management in Asia. Journal of Material Cycles and Waste Management, 15(4), 411–419.

[2] Puckett, J; Smith, T. (2002) Exporting harm: the high-tech trashing of Asia. The Basel Action Network. Seattle 7 Silicon Valley Toxics Coalition.

[3] Hino, T.; Agawa, R.; Moriya, Y.; Nishida, M.; Tsugita, Y.; Araki, T. Techniques to separate metal from waste printed circuit boards from discarded personal computers. J. Mater. Cycles Waste Manag. 2009, 11, 42–54.

[4] ME Emetere, H Jumbo, (2019). Computational extrapolation of municipal waste generation in Nigeria, Journal of Physics: Conference Series 1299 (1), 012032

[5] S.E. Musson, Y-C. Jang, T.G. Townsend and I-H. Chung. "Characterization of lead leachability from cathode ray tubes using the toxicity characterization leaching procedure". Environmental Science and Technology 2000, 34 pp. 4376-4381.

[6] Smith, D.; Small, M.; Dodds, R., Amagai, S. Strong, T. (1996) "Computer monitor recycling: A case study". Eng. Sci. edu. J., 5(4), 159 – 164, 1996.

[7] Waldir A. Bizzo, Renata A. Figueiredo and Valdelis F. de Andrade, (2014), Characterization of Printed Circuit Boards for Metal and Energy Recovery after Milling and Mechanical Separation, Materials 7: 4555-4566

[8] Pan, J.; Liu, Z.; Liu, G.; Wang, S.; Huang, H. Recycling process assessment of mechanical recycling of printed circuit board. J. Cent. South Univ. Technol. 2005, 12, 157–161.

[9] Fatta, D., Papadopoulos, A., and Loizidou, M.: A study on the landfill leachate and its impact on the groundwater quality of the greater area, Environ. Geochem. Hlth., 21, 175–190, 1999

[10] Satarug, S., Garrett, S.H., Sens, M.A., Sens, D.A., 2010. Cadmium, environmental exposure, and health outcomes. Environ. Health Perspect. 118:182–190. https://doi.org/10.1289/ehp.0901234.

[11] EU (2003). Directive 2002/95/EC of the European Parliament and of the Council of 27 January 2003 on the restriction of the use of certain hazardous substances in electrical and electronic equipment, Official Journal L 037 , 13/02/2003 P. 0019 – 0023

[12] Olubanjo, Kehinde, Osibanjo, Oladele, Nnorom Innocent Chidi, (2015), Evaluation of Pb and Cu contents of selected component parts of waste personal computers, J. Appl. Sci. Environ. Manage. 19 (3): 470 – 477

[13] Yoo, J.M.; Jeong, J.; Yoo, K.; Lee, J.C.; Kim, W. Enrichment of the metallic components from waste printed circuit boards by a mechanical separation process using a stamp mill. Waste Manag. 2009, 29, 1132–1137.

[14] IAEA (2001) Safety Report Series 19 STI/PUB/1102.

[15] IPSN, (2006), radioecology, https://www.irsn.fr/EN/publications/thematic/Documents/booklet_radioecology.pdf (Accessed 30/03/20

[16] CDC (Centers for Disease Control and Prevention), (2012). CDC response to advisory committee on childhood lead poisoning prevention recommendations in "Low level lead exposure harms children: a renewed call of primary prevention". U.S. Department of Health and Human Services, Atlanta, GA. http://www.cdc.gov/nceh/lead/ACCLPP/ blood_lead_levels.htm, (Accessed on 28/03/2020)

Acknowledgments
The authors acknowledge the partial sponsorship of Covenant University.

IOP Conf. Series: Earth and Environmental Science **633** (2021) 012018 doi:10.1088/1755-1315/633/1/012018

Parameter identification of capacitive transfer system based on Fitting Algorithm

Chen Chen[1,2], Chenhui Li[1,2] and Chenglin Liao[2]

1 University of Chinese Academy of Sciences, Beijing, China
2 Key Laboratory of Power Electronics and Electrical Drives, Institute of Electrical Engineering, Chinese Academy of Sciences
E-mail: kaizhiyu4014@gmail.com

Abstract. At present, in the wireless power transfer (WPT) of electric vehicle (EV), inductive power transfer (IPT) is still the main method. However, capacitive power transfer (CPT) has the advantages of no eddy current loss of coupling mechanism and light weight, which make it become a new research hotspot in the charging application of EV. But CPT system has high working frequency and sensitive parameters, therefore, in practical application, it is necessary to identify important parameters such as coupling capacitance and load. But at present, there are few identification methods for parameters of CPT. In this paper, a multi-parameter identification algorithm is proposed for capacitive wireless charging technology. Based on Vector Fitting algorithm and network synthesis principle, multi-parameter identification is realized. But in order to be widely used in practice, some parameters in circuit model are considered as known constraint. This can ensure that the proposed algorithm can not only realize multi-parameter identification, but also ensure the identification accuracy of system parameters in practical application.

1. Introduction

Environmental pollution and energy shortage are two major problems facing the automobile industry at present. With the rapid growth of the world's automobiles, the impact of traditional fuel vehicles on the environment and dependence on petroleum resources are becoming more and more serious. On the contrary, the advantage of EV is that they are less polluted to the environment, and even can achieve zero pollution. It can not only make the car get rid of dependence on petroleum resources, but also reduce the air pollution of the car.

At present, WPT can be classified IPT [1] [2], CPT [3] [4]. IPT system has been developed for a long time, and scholars at home and abroad have invested a lot of research, so the technology has become more mature. Figure 1 is the system structure of the CPT system. Because the system uses metal plates to couple, the distance between two metal plates is difficult to increase due to the limitation of the coupling capacitance value. This leads to the CPT system has been limited to short-range, low-power application scenarios.

In 2015, researchers such as D.C. Ludois and Lu Fei began to improve the power of CPT system. The designed system can be used to charge electric vehicles, and the system efficiency can reach 2.4kW power is 90.8% [5].

Unlike IPT system, CPT system has lower transmission loss in the environment with metal objects, and it can also resist the interference of magnetic field [6]. In addition, the coupling mechanism of CPT system uses metal plates, which not only makes the cost of system equipment lower, but also makes the weight of the whole system smaller.

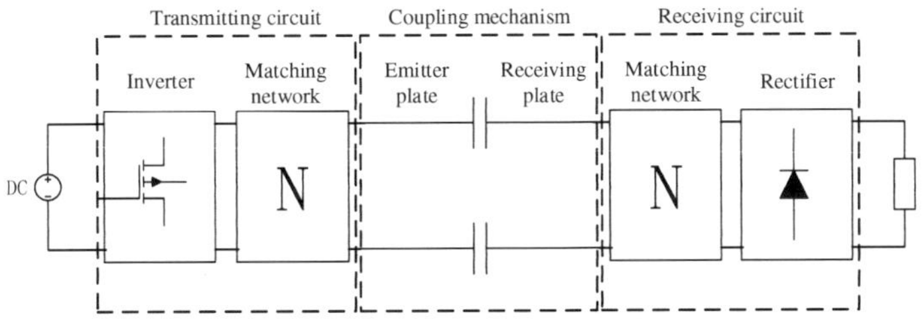

Figure 1. The CPT system structure.

When CPT system works, (1) various factors such as the placement of the metal plate of the coupling capacitor and its environment will lead to the system working under different conditions; (2) the system also guarantees that the system can work normally under the load of different parameters. It can be seen that the important parameters such as coupling coefficient and load in the actual system are not constant values, but constantly changing, which brings identification requirements to those parameters. There are some parameter identification methods in IPT system, but there are few parameter identification methods in CPT system [7].

In reference [8], the matching circuit model of IPT system is analyzed, and the relationship between input voltage and current and load resistance is deduced, so that the parameters of load resistance can be identified.

Reference [9] adopts step-by-step operation. Firstly, the standard load resistance is used to identify the coupling inductance, and then the actual load resistance is identified by the value of the previous step. By using this method, the multi-parameter identification of the system is realized.

However, in the field of CPT system, the reason for the lack of parameter identification methods is that the working frequency of capacitive wireless charging system is relatively high, and the resonant peak is relatively narrow, which makes the working state of the system vulnerable to the influence of system parameters.

However, in the field of CPT system, the reason for the lack of parameter identification methods is that the working frequency of capacitive wireless charging system is relatively high, and the resonant peak is relatively narrow, which makes the working state of the system vulnerable to the influence of system parameters. Once the parameters of the circuit model can not be accurately measured, the identification algorithm of IPT system described above will not get accurate results. Therefore, it is very important to study a more accurate algorithm to identify the parameters of CPT system.

2. Parameter identification method based on spectrum information

Figure 2 is the circuit model of CPT system. When the system load is stable, the system can be used as a single port network, and the components of the system are linear. Therefore, the CPT system can be regarded as a single-port network. By identifying the parameters of the single-port network, the parameters of the CPT system can be obtained. For a linear single-port network, its external characteristics can be determined by the driving point function of the port. Taking admittance function as an example, it is defined as follows:

$$Y(\mathrm{s}) = \frac{I_{in}(s)}{U_{in}(s)} \tag{1}$$

where, $I_{in}(s)$ and $U_{in}(s)$ are the voltage and current of the input end at different frequencies.

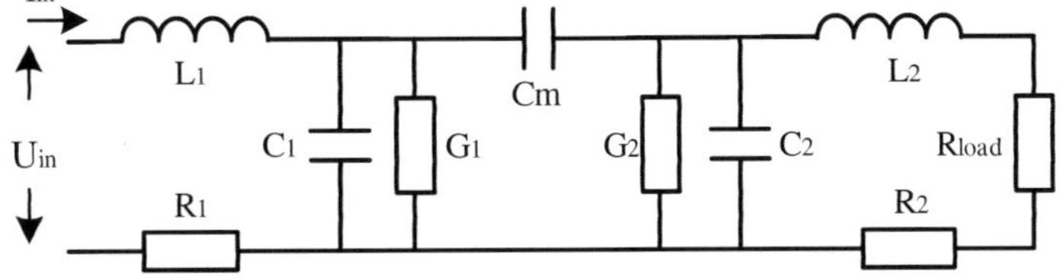

Figure 2. Circuit model of CPT system.

Supplying power to the system shown in figure 2, the corresponding admittance value is obtained according to equation (1) by measuring the corresponding voltage and current signals. We can constantly change the frequency of the inverters, so a series of admittance values are obtained. By fitting these admittance values with the fitting algorithm, the expression of the driving-point function of the circuit can be obtained.

By synthesizing the network parameters and combining the circuit model parameters in figure 2, the unique analytic expression of the system function can be obtained, and then combined with the fitting expression, the unknown parameters of the system can be calculated effectively.

Therefore, it is clear to identify the parameters of CPT system, which mainly includes the following contents: (1) the fitting of the driving-point function; (2) the synthesis of network parameters.

2.1. Fitting of driving-point function

For a single-port linear network, the driving-point function can be expressed as a rational fraction:

$$Y(s) = \frac{N(s)}{D(s)} = \frac{\sum a_i s^i}{\sum b_j s^j} \tag{2}$$

where, a and b in the above equation are all real numbers, and the order of the numerator and denominator is determined by the topology and component parameters of the circuit.

2.2. Fitting of driving-point function

In the process of fitting, it is generally assumed that the noise distribution of sampling values obeys the normal distribution rate, then the sum of squares and the minimum of fitting errors can be taken as the index, but the fitting function is a non-linear function, so the non-linear least squares is chosen for fitting:

$$\min \sum \left\| \tilde{Y}(s) - Y(s) \right\|_2 \tag{3}$$

where, $\tilde{Y}(s) = \frac{\sum_i \tilde{a}_i s^i}{\sum_j \tilde{a}_j s^j}$, $\tilde{a}_i$ and $\tilde{a}_j$ are undetermined parameters of functions.

If we directly using the sum of squared errors as an optimization object, the optimization problem is non-convex, there are multiple local optimum of the fitting error in the solution space. Therefore, if we use Levenberg-Marquardt method or the trust region method directly, it is difficult to converge to the optimal value for nonlinear optimization. In order to solve the local optimum problem of the fitting algorithm of the non-linear least squares method in rational fraction fitting, Sanathanan and Korener proposed the S-K iteration method of iterative solution, which transformed the non-linear least squares problem into the iteration process of the linear least squares problem[10], which can convert equation (3) to equation (4).

$$\min \sum \frac{1}{\sum_j \tilde{a}_j s^j} \left\| \sum_i a_i s^i - Y(s) \sum_j a_j s^j \right\|_2 \tag{4}$$

Taking LC series resonant circuit model as an example, we use S-K iteration method to linearize the objective function in the search space. The value of the objective function is shown in the blue surface in figure 3. Compared with the square sum error represented by the purple surface in the graph, the linearization function of S-K iteration method keeps the value of the objective function of the optimal solution position unchanged, while the value of the objective function outside the optimal point becomes flat and has a unique minimum value, so it can converge to the optimal solution better.

2.3. Fitting of driving-point function

In the fitting objective function expression (3) of the S-K iteration method, when the frequency range of sampling points is very wide, because of the existence of the high order term of s, the order of magnitude of the coefficients is very different in the linear least squares problem. Finally, the number of matrix conditions is very large, and the numerical accuracy is poor when solving the problem. To solve this problem, Bjørn and Adam proposed Vector Fitting algorithm [11], which reduces the condition number of matrix by transforming the basis of polynomial.

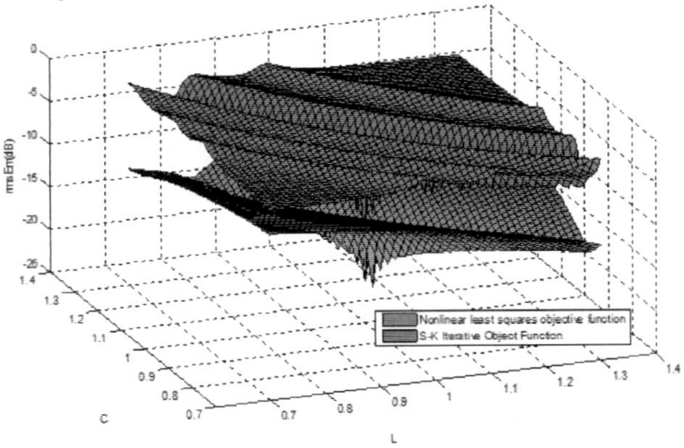

Figure 3. Linearization of objective function by S-K iteration method.

By transforming Equation (2) into zero-pole point through partial fractional expansion, the following fitting expressions can be obtained:

$$Y(s)=\sum_i \frac{r_i}{s-p_i}=\frac{\prod_i (s-z_i)}{\prod_j (s-p_j)} \tag{5}$$

Linearizing equation (3) and multiplying weight $\sigma(s)=\frac{\prod_j (s-p_j)}{\prod_j (s-\tilde{p}_j)}$, the linearized least squares problem can be obtained:

$$\min \sum \left\| \sigma\tilde{Y}(s)-Y(s)\sigma(s) \right\|_2 \tag{6}$$

Where

$$(\sigma\tilde{Y})(s)=\frac{\prod_i (s-z_i)}{\prod_j (s-\tilde{p}_j)}=\sum_i \frac{r_i^{(1)}}{s-\tilde{p}_i} \tag{7}$$

$$\sigma(s)=\frac{\prod_j (s-p_j)}{\prod_j (s-\tilde{p}_j)}=\sum_i \frac{r_i^{(2)}}{s-\tilde{p}_i}+1 \tag{8}$$

Because there is no higher order term of s in the expression, the number of conditions of each matrix of the linear least squares problem is smaller, so it can have better numerical accuracy in solving the problem.

The specific algorithm flow is as follows:

(1) Given the initial pole value of iteration $\tilde{p}_i$;

(2) Using equation (6), we can get $r_i^{(1)}$, $r_i^{(2)}$;

(3) By equation (8), p_i is solved by $r_i^{(2)}$ and $\tilde{p}_i$ as new pole.

(4) Then we use equation (6) to get $r_i^{(1)}$, $r_i^{(2)}$;

(5) To determine whether the fitting error of the algorithm is greater than the set limit, if it is greater than the set limit, the new solution will be recalculated as a new estimation $\tilde{p}_i=p_i$, back to (2). Otherwise, the value of p_i, $r_i^{(1)}$ and $r_i^{(2)}$ are directly used as fitting results.

2.4. Fitting of driving-point function

After using Vector Fitting algorithm to fit the driving-point function of the system, it is necessary to obtain the actual circuit parameters from the expression of the function, which is the synthesis of the

IOP Conf. Series: Earth and Environmental Science **633** (2021) 012018 doi:10.1088/1755-1315/633/1/012018

circuit network. In CPT system, Cauer synthesis method can be used because the topological structure of matching circuit is mostly T-shaped network structure [12].

For double-sided LC-matching CPT system shown in figure 2, the expression of input admittance function is obtained by Cauer synthesis method:

$$Y(s)=\cfrac{1}{L_1s+R_1+\cfrac{1}{C_1s+G_1+\cfrac{1}{\cfrac{1}{C_ms}+R_m+\cfrac{1}{C_2s+G_2+\cfrac{1}{L_2s+R_2+R_{load}}}}}} \tag{9}$$

We can know that the highest denominator order of admittance function is 4 after expand the formula obtained above:

$$Y(s)=\frac{a_3s^3+a_2s^2+a_1s+a_0}{b_4s^4+b_3s^3+b_2s^2+b_1s+b_0} \tag{10}$$

Therefore, the input admittance function is fitted by using the fourth-order Vector Fitting algorithm, and the circuit is solved by using equation (10). When network synthesis is used, the impedance of the coupling capacitor is very large in the CPT system, which leads to the loss resistance can not be identified properly. Therefore, we neglect the loss resistance, but there is a degree of freedom in the circuit. Because there is a degree of freedom in the circuit that can not be solved, it is necessary to know a receiver end parameter as a known condition to completely solve the circuit parameters [13]. In this paper, the inductance of matching circuit is taken as the known condition. The identification results shown in this section can be obtained. The result is that the parameter identification of the whole circuit is basically close to the model parameters.

Table 1. Noise-free identification results of parameter identification algorithm based on vector fitting.

Circuit parameters	Vector Fitting identification results	Model parameters
L1(uH)	82.24	82.24
R1(Ω)	3.4772	3.44
C1(pF)	288	288
Cm(pF)	19.998	20
C2(pF)	288.01	288
L2(uH)	82.24	82.24
R2(Ω)	3.3562	3.44
Rload (Ω)	33.4222	33.1

3. Vector fitting algorithm with constraint

The Vector Fitting algorithm introduced in the second section has no constraints, so we call it the unconstrained Vector Fitting algorithm (Unconstrained-VF algorithm). It can be proved that the algorithm can identify all the parameters in the circuit, however, in order to improve the applicability of the parameter identification, we consider the part of the component parameters in the model circuit which are not easy to change as the known conditions. The purpose of that is:

- In practical use, not all the parameters of components need to be identified, and some of them can be measured in advance.

- The Vector Fitting algorithm can be constrained by taking some component parameters as known conditions, which can improve the accuracy of parameter identification in practical use.

Some known component parameters divide the double-sided LC matching capacitive system shown in figure 2 into several parts. As shown in figure 4, the whole circuit model is divided into known and unknown parts. We can solve the problem by following steps:

Step 1: give initial values to the elements of the unknown part of the middle part and the rightmost part. Just like the coupling capacitance C_m and the load resistance R_{load}. We can take $\tilde{C}_m$ and $\tilde{R}_{load}$ as the initial values of C_m and R_{load}.

Step 2: at this time, it is reasonable to assume that the unknown part on the right (the load resistance $\tilde{R}_{load}$) becomes a known condition. When both sides are known, Vector Fitting algorithm is used to solve the parameters of the unknown part in the middle (the coupling capacitance C_m), and the parameters obtained are taken as known conditions. In turn, Vector Fitting algorithm is used to solve the unknown parameters on the right-most part (the load resistance R_{load}).

Step 3: through this continuous iteration method, until the final fitting error is less than the error limit, the final parameters can be used as identification results.

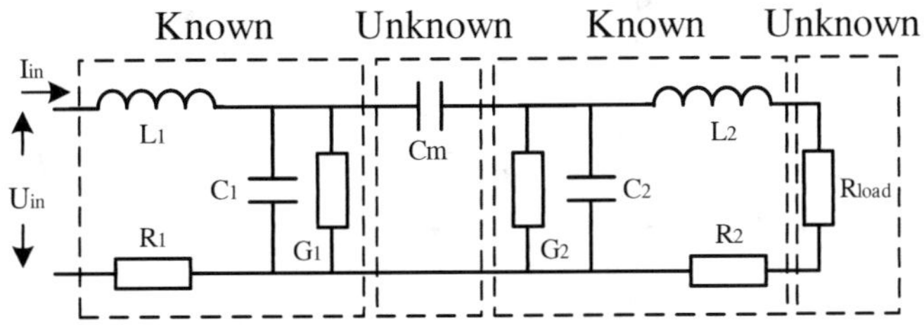

Figure 4. CPT circuit model segment.

This method is called Vector Fitting algorithm with constraints (Constrained-VF algorithm), if you want to identify more parameters, you can choose to divide the model into more parts, which are either known or unknown, then repeatedly use the Constrained-VF algorithm.

We then use the 2.2 section of the simulation model and the sampling parameters. However, we only identify the coupling capacitance C_m and the load resistance R_{load}. By comparing the simulation results with table 1, we can find that the Constrained-VF algorithm avoids the static error of the model caused by the loss resistance neglected in network synthesis. Table 2 only show the identified parameters C_m and R_{load}, others are all known parameters.

Table 2. Noise-free identification results of parameter identification algorithm based on constrained-VF algorithm.

Circuit parameters	Constrained-VF identification results	Model parameters
L1(uH)	-	82.24
R1(Ω)	-	3.44
C1(pF)	-	288
Cm(pF)	20	20
C2(pF)	-	288
L2(uH)	-	82.24
R2(Ω)	-	3.44
Rload (Ω)	33.121	33.1

4. Experimental verification

According to the circuit model of the double-sided LC matching shown in figure 2, we can make an actual circuit shown in figure 5. The inductance of the matching circuit part of the circuit is wound by the Litz wire, and the capacitance of the matching circuit part is formed by connecting a plurality of film capacitors having a capacitance of 1nF in series. The coupling capacitor is a 20pF ceramic capacitor. Since the determinism of the algorithm needs to be compared by the true value of the component, the individual components in the actual circuit are measured as true using the Agilent E4980A LCR meter.

Figure 5. Double-sided LC matching CPT system circuit physical map.

In the experiment, the inductance, resistance and capacitance of the matching circuit are taken as known conditions, and the identification of coupling capacitance and load resistance is selected. The true value of coupling capacitance and load resistance are 20 pF and 20 Ω, respectively. All the results are recorded in table 3, from which we can know that the identification results of the Unconstrained-VF algorithm and the Constrained-VF algorithm for the coupling capacitance are 19.429 pF and 20.5 pF, respectively, and the identification results of the load resistance are 24.2609 Ω and 20.6463 Ω, respectively. Compared with the results of load resistance identification, the identification accuracy of the constrained model is less than 1 Ω, while that of the unconstrained model is about 4 Ω, its identification error is about four times that of the constrained model.

Table 3. Parameter identification effect of parameter identification algorithm on model physical circuit.

Circuit parameters	Unconstrained-VF identification results	Constrained-VF identification results	Component measurement
L1(uH)	79.646	-	77.1773
r1(Ω)	2.8533	-	3.6928
C1(pF)	256.53	-	265.7394
Cm(pF)	19.429	20.5	20
C2(pF)	250.15	-	252.2711
L2(uH)	79.342	-	78.8456
r2(Ω)	2.5634	-	3.43
Rload(Ω)	24.2609	20.6463	20

The fitting curves of the two algorithms are shown in figure 6. Due to the narrower resonance peak of the CPT system, the fitting results of the two algorithms are quite different from the real curve in the two peaks. If we want to improve the identification accuracy of the algorithm, we can consider adding sampling points in two peaks, but it may be that too many sampling points lead to over-fitting, and the identification accuracy of the algorithm will decrease.

By comparing the two algorithms, it can be found that for the coupled inductor, both algorithms can get accurate conclusions, while the inverse of the load resistance, the results of the two algorithms have large deviations. Therefore, it is necessary to verify the identification range of the load resistance. In the experiment, by changing the resistance of the load resistance, two kinds of algorithms are used to identify the resistance of the load resistor. The result is shown in figure 7. The result shows that if the load resistance is in the range of 2 Ω to 200 Ω, the identification accuracy of Unconstrained-VF algorithm (calculated using equation (11)) is between 14.78% and 37.45%, but the identification accuracy of Constrained-VF algorithm is between 0.05% and 8.85%.

$$Y(s)=\frac{a_3s^3+a_2s^2+a_1s+a_0}{b_4s^4+b_3s^3+b_2s^2+b_1s+b_0} \tag{11}$$

where, $\sigma(\%)$ is the identification accuracy, $V_{fitting}$ is the resistance value obtained by fitting, $V_{ground\ true}$ is the ground true.

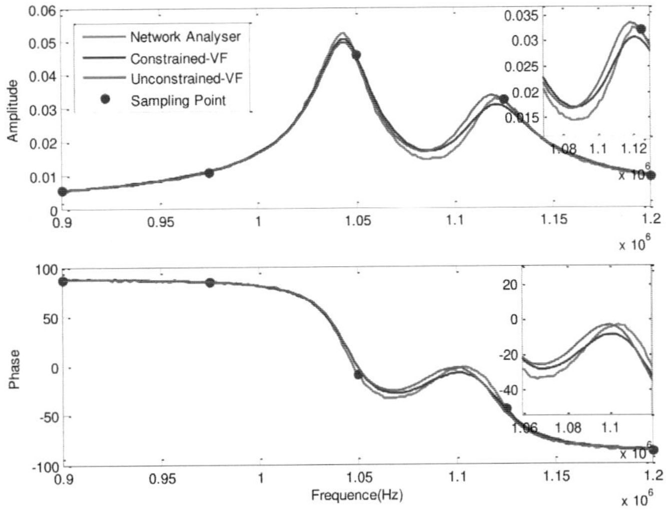

Figure 6. Comparison of fitting results of two kinds of parameter identification.

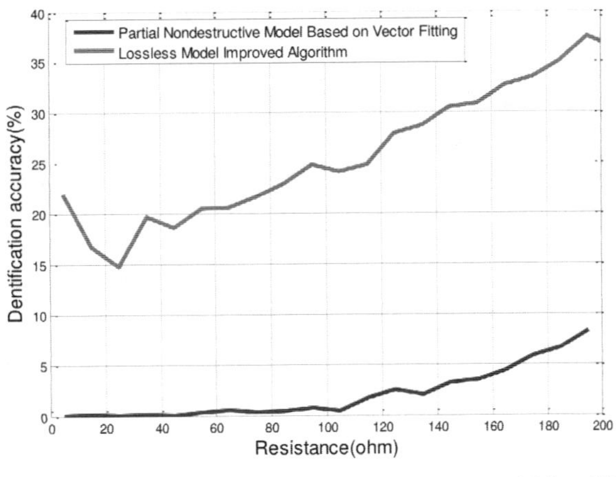

Figure 7. Relationship between load resistance change and identification accuracy.

5. Conclusion

CPT system have high operating frequencies and narrow resonant peaks, which makes them more sensitive to system frequencies, and system efficiency is also susceptible, which puts higher demands on system parameters. Therefore, we need to improve the parameter identification accuracy of the system to ensure efficient operation of the CPT system under weak coupling. In this paper, a parameter identification algorithm based on Vector Fitting is proposed, and the multi-parameter identification of the system is realized by combining the network parameter synthesis principle, as we called Unconstrained-VF algorithm. However, in order to ensure the applicability of the algorithm, some of the component parameters are selected as known conditions to constrain the Unconstrained-

VF algorithm, which we called it Constrained-VF algorithm. Finally, both of the above algorithm are verified for the doubled-sided LC-matching CPT circuit. The results show that the proposed algorithm can obtain higher-precision parameter identification while multi-parameter identification.

6. References

[1] A J. T. Boys and G. A. Covic, "The Inductive Power Transfer Story at the University of Auckland," IEEE Circuits and Systems Magazine, vol. 15, no. 2, May 2015, pp. 6-27, doi: 10.1109/MCAS.2015.2418972.

[2] L Pugi, A Reatti, and F Corti. "Application of modal analysis methods to the design of wireless power transfer systems," Meccanica. vol. 54, Jan. 2019, pp. 321-331, doi: 10.1007/s11012-018-00940-x.

[3] J. Dai and D. C. Ludois, "A Survey of Wireless Power Transfer and a Critical Comparison of Inductive and Capacitive Coupling for Small Gap Applications," IEEE Transactions on Power Electronics, vol. 30, no. 11, pp. 6017-6029, Nov. 2015, doi: 10.1109/TPEL.2015.2415253.

[4] C. Li, X. Zhao, C. Liao and L. Wang, "A graphical analysis on compensation designs of large-gap CPT systems for EV charging applications," CES Transactions on Electrical Machines and Systems, vol. 2, no. 2, pp. 232-242, June 2018, doi: 10.30941/CESTEMS.2018.00029.

[5] F. Lu, H. Zhang, H. Hofmann and C. Mi, "A Double-Sided LCLC-Compensated Capacitive Power Transfer System for Electric Vehicle Charging," IEEE Transactions on Power Electronics, vol. 30, no. 11, pp. 6011-6014, Nov. 2015, doi: 10.1109/TPEL.2015.2446891.

[6] C.Y. Xia, C.W. Li and J. Zhang, "Analysis of power transfer characteristic of capacitive power transfer system and inductively coupled power transfer system," 2011 International Conference on Mechatronic Science, Electric Engineering and Computer (MEC), Jilin, 2011, pp. 1281-1285, doi: 10.1109/MEC.2011.6025703.

[7] H.Y. Zhang. " Load and mutual inductance identification method of ICPT system based on switching capacitors," Chongqing University, 2015.nference on Mechatronic Science, Electric Engineering and Computer (MEC), Jilin, 2011, pp. 1281-1285, doi: 10.1109/MEC.2011.6025703.

[8] X. Dai, Y. Sun, C. Tang, Z. Wang, Y. Su and Y. Li, "Dynamic parameters identification method for inductively coupled power transfer system," 2010 IEEE International Conference on Sustainable Energy Technologies (ICSET), Kandy, 2010, pp. 1-5, doi: 10.1109/ICSET.2010.5684445.

[9] Z. Wang, Y. Li, Y. Sun, C. Tang and X. Lv, "Load Detection Model of Voltage-Fed Inductive Power Transfer System," IEEE Transactions on Power Electronics, vol. 28, no. 11, pp. 5233-5243, Nov. 2013, doi: 10.1109/TPEL.2013.2243756.

[10] C. Sanathanan and J. Koerner, "Transfer function synthesis as a ratio of two complex polynomials," IEEE Transactions on Automatic Control, vol. 8, no. 1, pp. 56-58, Jan. 1963, doi: 10.1109/TAC.1963.1105517.

[11] B. Gustavsen and A. Semlyen, "Rational approximation of frequency domain responses by vector fitting," IEEE Transactions on Power Delivery, vol. 14, no. 3, pp. 1052-1061, July 1999, doi: 10.1109/61.772353.

[12] S.Q. Luo, F. Liu, and Z.G. Han. Principles of Network Synthesis, 2rd ed., Tongji University Press, 2009.

[13] C.H. Li. " Design of Capacitive Power Transfer System for Electric Vehicle Charging," University of Chinese Academy of Sciences, 2018.

AUTHOR INDEX

Institute of Physics
Dirac House, Temple Back
Bristol BS1 6BE UK

ISSN: 1755-1307
ISBN 978-1-7138-2946-1